SpringerBriefs in Fire

Series Editor

James A. Milke
Department of Fire Protection Engineering
University of Maryland
College Park, Maryland, USA

More information about this series at http://www.springer.com/series/10476

Rosalie Faith Wills • André Marshall

Development of a Cyber Physical System for Fire Safety

Springer

Rosalie Faith Wills
Arup
Washington, DC, USA

André Marshall
University of Maryland
College Park, MD, USA

ISSN 2193-6595 ISSN 2193-6609 (electronic)
SpringerBriefs in Fire
ISBN 978-3-319-47123-5 ISBN 978-3-319-47124-2 (eBook)
DOI 10.1007/978-3-319-47124-2

Library of Congress Control Number: 2016953851

Printed on acid-free paper

This Springer imprint is published by Springer Nature
The registered company is Springer International Publishing AG
The registered company address is: Gewerbestrasse 11, 6330 Cham, Switzerland

*I would like to dedicate this report
to my parents, Sarah and Joe Wills,
for their lasting love and support for
me and for my aspirations.*

Acknowledgements

The Chief Donald J. Burns Memorial Grant provided by the Society of Fire Protection Engineers and funded by Bentley Systems supported this research. Siemens provided equipment and resources such as a customized Fire Alarm Control Panel. The Maryland Fire and Rescue Institute provided resources and facilities to accomplish full-scale fire testing. Dr. Darryl Pines, Dean of the Clark School of Engineering, provided seed funding in support of this research.

I would like to thank my thesis advisor, Professor André Marshall, for his support and guidance throughout the duration of this project. I appreciate his ability to envision the future goals of the project while also providing clear guidance on current specific tasks.

I would also like to thank Professor Arnaud Trouvé and Professor Stanislav Stoliarov for being on my committee and providing me valuable guidance.

None of this would have been possible without the support of Professor Jim Milke for the past 5 years. I sincerely thank him for his genuine encouragement, insightful advice, and great sense of humor.

Contents

List of Figures

List of Tables

Chapter 1
Introduction

A cyber physical system (CPS) is a system of collaborating computational elements that monitor and control physical entities. In this project, a CPS system was developed as a test bed to explore next-generation fire safety. The CPS developed in this project utilizes current technologies in the modern built environment and emerging virtualization concepts. Sustainability goals, security concerns, and rapidly evolving information technology have driven a profound expansion in the use of sensors in the modern built environment. This rich sensor data, typically used for building services related to comfort, security, and energy management, can be integrated with fire sensor information to inform emergency decisions in the event of a fire.

The smart networked fire test bed developed in this study consists of a multistory fully instrumented building, a variety of fire and non-fire networked sensors, and a computational framework anchored by a Building Information Modeling (BIM) representation of the building. Building Information Modeling has become an increasingly popular design tool allowing virtual construction of the building supporting solution of physical construction and maintenance design challenges. In the test bed established for this study, well-controlled full-scale fire experiments are conducted (physical elements) and represented in the three-dimensional (3D) BIM model (computational elements) allowing for visualization of critical static and dynamic building and fire information needed to support fire fighter decisions. The literature review focuses on different aspects of CPSs' and how they contribute to the fire problem.

1.1 Motivation

Over the years, changes in modern infrastructure have introduced new challenges to fire fighters. Training and research programs have been developed to manage these challenges but there are still significant losses from fires each year.

© The Author(s) 2016   1
R.F. Wills, A. Marshall, *Development of a Cyber Physical System for Fire
Safety*, SpringerBriefs in Fire, DOI 10.1007/978-3-319-47124-2_1

This project focuses on both the fire safety challenges and opportunities presented by modern infrastructure. These opportunities include harvesting the environmental data collected by intelligent infrastructures to improve situational awareness in a possible fire environment. Another increasingly popular technology, BIM, presents a major opportunity for not only harvesting physical data, but also visualizing physical and environmental data for fire scene evaluation and decision making (i.e. size up). The advancements in emerging smart sensor technology and more sophisticated models have generated a new area of technology called CPS. This project focuses on utilizing these advancements to create a CPS for fire safety.

1.1.1   Modern Fire Safety Challenges

Modern fire fighting has evolved with advancing technology, improved training, and new research. Even with these developments, the consequences of fire continue to be tremendous. U.S. municipal fire departments responded to an estimated 1,375,000 fires in 2012. These fires killed 2855 civilians (non-firefighters) and caused 16,500 reported civilian fire injuries. Direct property damage was estimated at $12.4 billion dollars. Sixty-four firefighters died while on duty or of injuries incurred while on duty. The 480,500 structure fires accounted for 35 % of all reported fires [1].

The current research project was funded in part by the Chief Donald J. Burns Memorial Grant, supported by Bentley Systems, Incorporated. This grant was created to advance fire fighter science and technology. Chief Burns died in the collapse of the World Trade Center Towers on September 11, 2001, while setting up his command post to direct the evacuation. The purpose of this grant is to advance information modeling as a means of improving infrastructure safety and first-responder preparedness. To achieve this goal, it is important to understand current methods used by first-responders and the challenges presented by the modern built environment.

Fire fighters currently make tactical decisions with limited information based on quick, visual assessments of the emergency. 'Size-up' is a strategy used by fire fighters through these assessments to determine information like the building's height and area, class of construction, occupancy type, the location and extent of the fire, as well as any exposure concerns. Determining this key information efficiently is critical when fighting fires. Modern structure fires can develop so rapidly that seconds can change the destruction level significantly. New methods for fire safety need to be developed to respond to the challenges associated with the modern built environment. Changes in modern structures such as open layouts and economy construction (materials and methods) have resulted in new and unpredictable hazards, not the least of which is rapid fire growth. The emergence of CPS technology, with the potential to deliver real-time high quality actionable information to fire fighters, presents opportunities to address these extreme challenges in ways never before possible. A modern fire-fighting framework is envisioned where mission critical information is acquired and visually communicated in a CPS supporting informed, fast, and effective fire fighter decisions and actions.

1.1.2 Modern Infrastructure and Technology Opportunities

A sustainable infrastructure is designed to allow humans and nature to coexist in productive harmony environmentally, economically and socially. Resources that are needed by both parties for survival and well-being are utilized efficiently. Sustainable goals are driving modern infrastructure in a direction that causes new issues for fire protection. These goals also bring benefits for fire protection that should be utilized. The increased sensor density and building intelligence (i.e. CPS) used in modern infrastructure deserves the attention of the fire safety community with a focus on new opportunities to advance fire emergency response effectiveness and safety.

Sensing is a key function of intelligent building infrastructure, and therefore a significant amount of research has examined various sensing modalities and techniques. Smart wireless sensor networks have emerged as enablers for delivering sensor data. There are many sensors that are currently being used in these new sustainable structures. Temperature, light, and humidity are some of the aspects that are monitored. Using this data, building automation systems manage resources within the building. For example, an occupancy sensor monitors motion within a room and sends a signal to the automation system if the motion state changes and detects an occupant in the room. The automation system will then activate the lighting system to provide light to the room and the occupant. Intelligence can also be added to the system when combined with a contact sensor on a door to the room. The state of the door provides additional information to determine if an occupant is actually within the room. The combination of multiple types of sensors communicating together through an automation system efficiently and intelligently utilizes the energy required to maintain occupant comfort. Occupancy is just one environmental condition that is monitored and affects the building automation systems and building utilities. A list of some of the sensors and building utilities that would be in a modern sustainable structure is shown in Fig. 1.1.

Not only has the variety of sensors increased, but also the sensor's capabilities. All types of sensors, such as fire, security, energy, and others have all significantly advanced. In 2008 The Society of Fire Protection Engineers (SFPE) released an article "Factors in Performance-Based Design of Facility Fire Protection" [3]. It discusses how fire alarm system technologies have become far more advanced, with simpler circuits and more capabilities. The devices themselves are smart and can tell the panel when something is wrong. For example, a smoke detector knows when it

Fire Protection	Occupant Comfort	Security
Smoke Detectors	Thermostats	Occupancy Sensors
Heat Detectors	Humidistats	Contact Sensors
Valve Monitors	Lighting	Intrusion Detectors
Fire Alarm Control Panel	Heating/Cooling	Cameras
Fire Alarm Annunciator Panel	Zone Control Panels	Access Management

Fig. 1.1 Sensors and utilities within a modern sustainable structure

FIRE ALARM ANNUNCIATOR PANEL

NORTH

LEGEND	
FACP	FIRE ALARM CONTROL PANEL
FARA	FIRE ALARM REMOTE ANNUNCIATOR
⊕	STANDPIPE LOCATION
▶	POINT OF ENTRY
——	SPRINKLER/FIRE ALARM ZONE
Y	FIRE DEPARTMENT SPRINKLER CONNECTION

- (G) INDICATES GREEN LED
- (R) INIDCATES RED LED
- (Y) INIDCATES YELLOW LED

NORMAL — SILENCE

AUTO — ON

SYSTEM RESET   FIRE PUMP REMOTE START

NORMAL — SILENCE

AUTO — ON

TROUBLE SILENCE   GENERATOR REMOTE START

AUTO UNLOCKED

(PUSH BUTTON)   MANUAL UNLOCKED

LAMP TEST   STAIR DOOR LOCKS

DEVICE

- (R) MANUAL STATION
- (R) SMOKE DETECTOR
- (R) HEAT DETECTOR
- (R) ELEVATOR LOBBY/MACHINE ROOM SMOKE DETECTOR
- (Y) DUCT MOUNTED SMOKE DETECTOR
- (R) SPRINKLER WATER FLOW
- (Y) STANDPIPE WATER FLOW
- (Y) FIRE SERVICE LINE VALVE TAMPER
- (Y) FIRE SERVICE LINE VALVE FLOW

- (Y) STAIR 1
- (Y) STAIR 2
- (Y) STAIR 3
- (Y) STAIR 4
- (Y) STAIR 5
- (Y) STAIR - RETAIL

EMERGENCY SYSTEM STATUS

- (G) POWER ON
- (R) SYSTEM ALARM
- (G) SYSTEM TROUBLE
- (Y) FIRE PUMP RUNNING
- (G) FIRE PUMP FAULT
- (Y) GENERATOR RUNNING
- (Y) GENERATOR FAULT
- (Y) SUPERVISORY

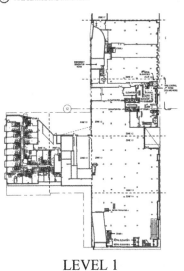

LEVEL 1

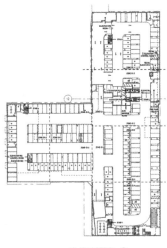

LEVEL 2

Fig. 1.2 Fire alarm annunciator

is dirty, adjusts its own sensitivity and reports this to the panel [3]. The report also states that these alarm systems can be programmed to display any information desired to responders. For example, the system can tell a responder where every device is located and what type of device it is [3]. Fire Alarm Light Emitting Diode (LED) graphic annunciators are used to show a fire alarm device location within a floor layout. This system can be useful but the two-dimensional representation on the fire alarm panel can be difficult to interpret and provides limited use until first-responders physically arrive on scene and locate the panel.

This project utilizes the information gathered from typical fire protection sensors such as smoke detectors, while also considering data from non-fire building sensors. In the same way that Global Positioning System (GPS) mapping has led to convergence of different kinds of information, from navigation tools to restaurant reviews, sensor integration can create a virtual framework offering a window into the otherwise invisible details and dimensions of a fire.

1.1.3 Cyber Physical Systems

A cyber physical system (CPS) is a system of collaborating computational elements that monitor and control physical entities. An example of a CPS tool is a Building Information Model (BIM). BIM is officially defined in the National BIM Standard (NBIMS-US™) as a digital representation of physical and functional characteristics of a facility [4]. Modern construction use BIMs to visualize a project in a three dimensional cyber physical environment before completing the project in reality.

A 2004 NIST study identified a lack of interoperability as an additional cost to the U.S Capital Facilities Industries of $15.8 billion annually [5]. The collaboration and exchange of information between the many stakeholders involved in the building process needs to be improved. Although other factors influence interoperability, BIMs can significantly aid this issue. The national BIM standard references this study and states that using BIM principles and practices, elements of the capital facilities industry are represented and exchanged digitally. Digital representation means that computers can be used to 'build' the capital facility project virtually, view and test it, revise it as necessary, and then output various reports and views for purchasing, fabrication, assembly and operations [5]. An example of a complex BIM created with Bentley AECOsim Building Designer is shown in Fig. 1.3 [6]. Three different views of the various disciplines are shown within the figure; the left view displays the realistic outside of the final building design, the middle view focuses on the structural elements, and the right focuses on the mechanical elements. BIMs have the capability to store a tremendous amount of information across multiple disciplines but can also be manipulated to only show specific information of interest.

BIMs have many uses to designers and project managers. The model itself can be used to visually communicate design concepts to the owner. A major utility of a BIM is clash detection, which describes how the model electronically checks to

Fig. 1.3 Multidiscipline features of BIM

make sure building systems are designed properly. Making changes electronically rather than on the job site reduces change orders, which minimizes the opportunity to jeopardize the project's budget and schedule [4]. Along with the many benefits already realized, it is also beneficial to utilize BIM as a tool for fire safety.

BIMs have the ability to store, manage, and visualize three-dimensional static and dynamic building and fire environment information. Static information elements that are included in modern BIMs that are also important for fire safety include but are not limited to; building geometry, construction materials, use, occupancy, water supply, hydrant locations, location of the annunciator panel, and location of the standpipes. This study considers these elements and also incorporates dynamic information such as room temperature, smoke obscuration, and carbon monoxide (CO) concentrations into a BIM. This study defines a framework that visualizes a fire environment a BIM and how it can be used to improve fire safety. This framework is based on a literature review of past fire safety research and a fire protection engineering assessment of building performance criteria.

1.2   Literature Review

Inverse fire models (IFMs) and building information models (BIMs) are emerging tools that can be brought together to provide a cyber physical system for fire safety. Traditional forward fire modeling uses initial conditions such as expected fire size, heat release rate (HRR), and room geometry to determine the impact of the fire on the environment. An inverse fire model uses the measurements within the fire environment and then calculates the fire size and fire location. The use of inverse fire models have recently become more relevant because the increased sensor density in the modern built environment. Sensors can provide the environmental conditions such as temperature profiles, ceiling jet temperatures, heat fluxes, optical densities, CO concentrations, and gas velocities. Using these measurements, and an IFM, a fire size and location can be calculated. Various studies have attempted to develop a reliable IFM with varying results. The accuracy of the IFMs is limited to known variables and computing capabilities. This section provides a review of the methods taken to determine fire size using compartment fire dynamics and IFMs.

1.2.1   Inverse Fire Model Development

Past studies have developed various inverse fire modeling methods to calculate fire information based on environment measurements. Ceiling jet temperatures are used by Lee and Lee as well as Davis and Forney to calculate the heat release rate (HRR) of a fire within a compartment [8–10]. Pairing an optimization algorithm focusing on the sequential regularization approach with the Fire Dynamics Simulator (FDS 81), Lee and Lee determined an estimate of the location and size of the fire using temperature measurements along the ceiling inside a compartment [8]. The research compared temperature profiles inside a compartment to predictions from an FDS model of the compartment. By minimizing the residuals between measured and predicted values, the location and heat release rate of the fire were calculated.

Davis and Forney developed the National Institute of Standards and Technology's Sensor-Drive Fire Model (SDFM) to calculate fire size [9, 10]. The SDFM uses ceiling-jet correlations with sensor measurements to approximate heat release rates (HRR) of fires. The HRRs are used as an input for a Consolidated Model of Fire and Smoke Transport (CFAST). CFAST, a computational inexpensive model, then makes quick predictions of HRR growth, hot gas layer temperatures, smoke/visibility concerns, and even predict structural failure. CFAST is a zone fire model that performs more simplified calculations than a full scale CFD model.

FireGrid explored a variety of methods aimed to perform super-real time modeling of building fires [11, 12]. The project used measurements from full-scale compartment fire experiments undertaken in Dalmarnock, Glasgow in 2006. A six-room experiment, the compartment was instrumented with temperature, heat flux, optical density, gas velocity and structural monitoring sensors. FireGrid also explored taking data from numerous building sensors to obtain temperature, smoke, and CO measurements to plug into a simultaneous Computational Fluid Dynamics (CFD) and Finite Element models. This approach was able to calculate not only fire size and location, but also the impact the fire had on the structure. Ultimately, the FireGrid approach was too computationally expensive to achieve its super-real time goals.

Neviackas developed an inverse fire model aimed to determine the heat release rate of a fire in a compartment given the temperature of a hot gas layer over a function of time [13]. The IFM was initially successful for a variety of configurations and fuel sizes within 10 % and on the order of 1 %. However, the IFM solution became non-unique when incorporating the size of doors and windows in any given room. Recently, Price conducted multi-room compartment fire experiments to obtain measurements of hot gas layer temperature and depth [14]. These measurements were used as an input to the inverse fire model created by Neviackas that coupled a genetic algorithm with a zone fire model to calculate a unique solution to the original fire size and door opening used in the experiments. The objective of this research was to calculate simultaneously the real-time fire size and fire door opening of the experiment using a combination of hot gas layer temperature and hot gas layer height measurements from a multi-room compartment in concert with an inverse fire model. This study emphasized the need for knowing the ventilation state of the fire in order to determine the fire size.

Many different approaches to calculating fire size in a building have been attempted. From these studies it is determined that temperature profiles, ceiling jet temperatures, optical densities, CO concentrations, and ventilation conditions are important environment conditions to measure in order to calculate fire size using an inverse model. The approach will utilize this information when determining what types of sensors to use for the full scale fire experiments.

1.2.2   BIM Utilization Concepts

The first phase of the Chief Donald J. Burns Memorial Research Grant was awarded to O.A. Ezekoye to explore the use of BIM for improving fire safety. Anderson and Ezekoye designed a framework to provide an informed risk analysis of a building using a combination of BIM software, fire models, and statistical analysis [15]. A fire scenario is created when a potential heat source overlaps with a fuel package, the hazard is identified and a fire scenario is created featuring the heat source as an ignition source on the object. A statistical analysis is applied to these features along with the other features that affect the development of the fire environment such as probability of vent opening area and probability of an alarm or sprinkler operation [15]. The end goal of the framework is to improve the level of fire hazard and risk information that is available to the building owner, fire service, or any other interested stakeholder. This project provided insight of how BIM can be utilized for forward modeling to predict fire scenarios. Ultimately, a fire event could be analyzed by coupling this framework that can predict fire scenarios with cyber physical framework of this study of an actual fire scenario. This process could provide insight into the size of the fire, growth rate of the fire, and forecast the future fire environment.

The Society of Fire Protection Engineers (SFPE) that supported the previous project also released a position statement in 2011 about BIM and fire protection engineering [16]. The statement discussed the benefits and challenges of integrating BIM and fire protection engineering design such as fire suppression systems design, fire alarm and notification systems design, life safety and code compliances, along with performance based design. There is potential to develop specific tools within the BIM platform to automatically incorporate system design and performance characteristics such as hydraulic and water supply calculations, atrium smoke control calculations, fire effects, and egress modeling. The review of the worth of combining BIM and fire protection engineering provides insight for the relevance of also using BIM as a method of displaying critical building and fire information.

The National Institute of Standards and Technology (NIST) conducted a workshop to identify information needs for emergency responders during building emergencies [17]. The goal of the workshop was to gain guidance on what specific building information would be of greatest benefit to public safety officials as well as how to best present it. A report of their findings provides insight for developing the cyber physical framework for this project. The information is broken down into two types, static and dynamic information. Static information is defined as the building information that is available before an incident. Examples of static information are given as room geometries, sensor locations, and ventilation system schematics. All of these items can be represented using BIM, although the study does not reference this tool.

Dynamic information is defined as the set of information that comes from real-time status of building system controllers and sensors. The report describes dynamic information as including both direct sensor readings and the output of decision support tools that analyze the sensor data [17]. Although many different types of data are referenced, two main examples are specifically given; one being a fire decision tool that would use the fire system sensors along with knowledge of the building geometry to produce fire size, location, and fire growth rate estimates. The other example is a security support tool that estimates occupant location based on occupancy sensor and light system reports. The first example refers back to the concept of an inverse fire model discussed in the previous section of the literature review. Although an inverse fire model is not mentioned within the report, with the right sensor measurements and the proper tools, the dynamic information such as the fire size can be determined. Both of these examples directly influenced the sensors chosen for this project.

The display of the static and dynamic information is explored during the NIST workshop. A three dimensional display was discussed, and is determined that in order to be useful the display must be both rich in information but presented in a simple uncluttered easy to use interface. Examples of icons were approved such as a person icon for an occupant in need of rescue, and a fire icon for the location of the fire. It was also noted that the attendees did not think that text only displays would work. This supports the need for the use of a three dimensional BIM model that can represent a vast amount of information but can also limit the display to what is specifically important.

Within the NIST report there is also an Appendix that provides an example scenario using a three dimensional program for an incident command tactical system. The Appendix provides visual examples and reasoning of how to display the various types of dynamic information. Two examples are shown in Fig. 1.4 [17]. The temperatures measured by a heat sensor in each room are given an associated color depending on the amount of heat in the room, as shown in the left of Fig. 1.4. Although exact temperature thresholds for each color were not determined, it was stated that red would best represent a hot temperature. The example also indicates that it is better if only areas that are given an associated color are those that register as abnormal. Location of people in need of rescue is one of the most important

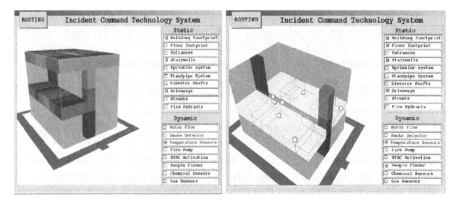

Fig. 1.4 Examples of incident command tactical system

pieces of information for a first responder. A visual on the location of the people is shown in the image on the right of Fig. 1.4.

In 2006, NIST expanded on the previous workshop and developed an outline for intelligent building response for safety officials [18]. A technology data path is proposed that will allow information collection and transport to the emergency response. A demonstration of the technology with a decision support system transmitting real-time building information to first responders was held at NIST.

The decision support system is a Sensor Driven Fire Model (SDFM) that converts sensor signals from the fire sensors to predictions usable by first responders. The proposed SDFM is a computer modeling software that has the capabilities to provide the following analysis; uses heat and smoke sensor signals to identify fire growth and size, provides hazardous condition warnings for first responders, uses sensor signals to identify open/closed door status to provide real-time building configuration information.

The model is planned to have the warnings that can be shown as a series of colors on a building floor plan. The warnings are based on smoke layer height measurements and temperature measurements. The first warning is for visibility problems when a smoke layer is measured 2.0 m above the floor. The second warning level occurs when the smoke layer is dense enough to be toxic and a breathing apparatus is needed. This occurs when the smoke layer has descended to 1.5 m above the floor and reached a temperature of 50 °C. The third warning is when the temperature is at 500 °C, when flashover can occur and the area needs to be evacuated immediately.

Guidance of how to color code these warnings are provided in Appendix C, the NIST Experimental Implementation Report [18]. The interface is shown as a single floor building plan with various colors indicating different levels of alarm in each compartment. These colors are represented in Fig. 1.5 and are the following; white is free of any hazard, blue indicates that a possible hazard has been detected, green indicates that a hazard has been detected, yellow is a significant hazard and that the compartment can only be entered using special equipment, and red means the compartment has flashed over and is lethal to anyone. The conclusions in these studies influence the approach for developing a CPS for fire safety.

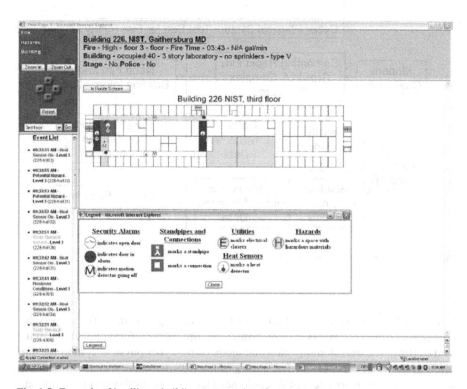

Fig. 1.5 Example of intelligent building response interface

1.3   Project Focus

1.3.1   *Objectives*

In this study a cyber physical system (CPS) framework was developed to

- Demonstrate real-time fire information viability through visualization of a measured fire environment
- Provide validation data for continued CPS development

1.3.2   *Scope*

The CPS framework was developed for an existing Maryland Fire and Rescue Institute (MFRI) training structure. The cyber element of this study utilizes Microstation, a Building Information Model (BIM) software created by Bentley Systems. This BIM is coupled with physical environmental information gathered by conventional fire and non-fire sensors and laboratory instruments during

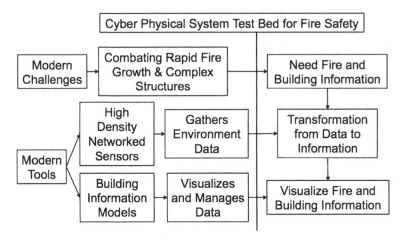

Fig. 1.6 Cyber physical system for fire safety

wellcontrolled full-scale fire experiments at the training structure. The data gathered during fire experiments is transformed into fire safety information and visualized in BIM. The overall motivation and methods for cyber physical system for fire safety is shown in Fig. 1.6. The framework of the CPS determines what information an emergency fire responder (EFR) could need and how to display that information. It is demonstrated that collaboration of sensor data and a BIM is the start of a CPS that is useful for EFRs to make more informed decisions in the event of an emergency. If EFRs could make more informed decisions, respond faster, and more effectively to the continuously and often rapidly changing hazards associated with modern fires, they could save more lives while minimizing their exposure to hazards and risk of injury.

Chapter 2
Approach

In this study a cyber physical system (CPS) framework was developed to demonstrate the viability of delivering real-time fire information for fire safety and to provide validation data for continued CPS development. The Maryland Fire and Rescue Institute (MFRI) training structure was the physical test bed for the CPS framework. In the physical environment the structure was fully instrumented with commercial sensors and experimental sensors and is used to conduct full-scale fire experiments in a complex geometry. The fire environment was controlled by well-characterized source and ventilation conditions defined in a test matrix. In the virtual environment the structure was characterized in a BIM by its static information such general building materials, geometries, occupancy classification, and use. The BIM and the data gathered in the experiments are coupled to demonstrate a CPS that has static information as well as dynamic information. Using guidance from past studies, a framework has been developed to display fire safety information that caters to the capabilities of the BIM and the measurements taken by the sensors.

2.1 Physical Environment

An objective of this project was to create, observe, and measure a physical fire environment through full-scale experiments. The experiments were conducted at a training structure at the Maryland Fire and Rescue Institute (MFRI). Full-scale fire experiments were conducted at the training structure with various fuel packages and ventilation characteristics. The experimental methods were predetermined by first creating the desired environment in a NIST tool, the Fire Dynamic Simulator (FDS) [7]. FDS is an open source freeware fire model that is the leading fire simulation software used by fire safety engineers and fire researchers. The desired environment was achieved when certain given ventilation and fuel conditions result in substantial environment changes but aren't so extreme that would exceed the capabilities of the training facility and instrumentation. The controlled desired environment safely

© The Author(s) 2016
R.F. Wills, A. Marshall, *Development of a Cyber Physical System for Fire Safety*, SpringerBriefs in Fire, DOI 10.1007/978-3-319-47124-2_2

Table 2.1 Testing matrix

Test #	Fire		Configuration		Sensors			
					Therm	Load cell	Smoke detector	Security
	Small (Crib)	Large (Pallet)	Door 2 Open	Door 2 Closed	Temp	Mass loss	CO, temp, smoke obs.	Contact, occupant
1	X		X		X	X	X	X
2	X			X	X	X	X	X
3		X	X		X		X	X
4		X		X	X		X	X

allows for the experiments to be measured at all stages of the fire from growth, peak, decay, and during ventilation until normal conditions are reached.

Each element of the testing matrix such as fuel type, fuel configuration, sensor location, ventilation conditions, etc. was evaluated. Past studies and research influenced the prescribed elements of testing. The testing matrix is shown in Table 2.1. The specific fuel packages and ventilation conditions provided comparably different scenarios in each testing case. The goal is to create an environment that is representative of a realistic fire scenario and will provide useful data for continued CPS development. Data gathered such as temperature, smoke concentration, carbon monoxide concentration, fuel mass loss, and ventilation conditions will provide validation data for future inverse fire models.

Specific Siemens sensors were installed into the training structure to monitor dynamic environmental conditions during the fire tests; multi-criteria smoke detectors monitor temperature, smoke obscuration, and carbon monoxide concentrations; contact sensors monitor the condition of windows (open/close); and occupancy sensors monitor if an occupant is present in the area. These measurements were communicated to a control panel and stored in data file. An overview of the sensors is shown in Table 2.2. Thermocouples measuring temperatures and a load cell measuring the mass of the fuel during the duration of the fire experiment provide accurate measurements of the fire. The information gathered by these sensors demonstrated a rich dataset of this dynamic information that is not only integrated into this study's CPS but can also be for more elaborate CPSs that may include data assimilation and inverse modeling features.

2.1.1 Testing Facility

The Maryland Fire and Rescue Institute (MFRI) of the University of Maryland is the State's comprehensive training and education system for emergency services. The institute plans, researches, develops, and delivers programs to enhance the ability of emergency services providers to protect life, the environment, and property. MFRI has had a long time standing partnership with the Department of Fire

Table 2.2 Sensor overview

	Siemens Desigo™ Fire Safety System				
	252 Point Fire Alarm Control Panel—Model FC 2025			Security Panel—Model D8125INV	
Sensor	Smoke detector			Contact sensor	Occupancy sensor
Model #	FDOOTC441			EN1210W	EN1260
Alarm type	Temperature	Smoke Obscuration	Carbon Monoxide	State of Window or Door	Presence of an occupant
Alarm threshold	57 °C–79 °C	3 % Obscuration/ft	30 ppm, 50 ppm	Open or closed	Yes or no
Direct measurement	Yes	Yes	No	No	No
Location	Each room, ceiling mounted			First floor	First, second, third floor
Number	13			7	3

Fig. 2.1 MFRI structural fire fighting building

Protection Engineering at the University of Maryland. This partnership encouraged the availability of MFRI facilities for use in this project. The facility of interest for this project is the MFRI structural fire fighting building, which is used as a training structure.

The four story MFRI structural fire fighting building is shown in Fig. 2.1. The footprint of the building is approximately 12.6 m by 7.8 m and each floor is approximately 2.8 m in height. The size and construction of the training facility allowed for full-scale fire scenarios to be designed and tested safety within the structure. The building is constructed with fire resistant materials to withstand extreme fire conditions created by fire fighter training exercises. Protective materials are applied to the walls, the doors and stairs are made of metal, and building utilities are limited. An extensive building description and photos of the protective materials and building elements are in the Appendix.

2.1.2 Sensor Selection and Instrumentation

Intelligent building infrastructure use sensors that monitor the environmental conditions to determine the specific need for building utilities. These sensors have become increasingly prevalent and can gather a high density of data. This project creates a cyber physical test bed that reflects these complex intelligent building infrastructures on a smaller scale. This cyber physical test bed was created by choosing the sensors that fulfill the fire safety goals and objectives of this project. Environmental sensors, security sensors, fire and smoke sensors, HVAC and airflow sensors are all considered. For this project the data that was focused on are measurements that not only provide information about the environment, but also can be used for fire safety such as determining a fire source. Past studies like the one by Price, determined that temperature, smoke layer heights, room geometries, and ventilation conditions can be used to determine the size and location of the fire [14]. These data types not only can be used to determine the characteristics of the fire, but also give a direct measurement that is useful for emergency first responders such as temperature. Ultimately the gathered data was used as dynamic information within the BIM using Microstation. The chosen sensors need to be compatible with Microstation in order to achieve this goal. The fire safety system designed for the training facility also needed to be designed to be portable, this process is explained further in the Appendix.

Siemens has had a long time standing partnership with the Department of Fire Protection Engineering at the University of Maryland. This partnership encouraged the availability of Siemens resources for use in this project. All Siemens sensors and data management systems are reviewed and specific ones are chosen for this project based on applicability and feasibility. Data sheets for all sensors and products used as well as product descriptions are provided in the Appendix. Siemens donated the chosen sensors and management systems to this project. Siemens provided consultation with how to use the products, designed and programmed the system, and provided in depth data sheets on how the products operated. An overview of the instrumentation is shown in Table 2.2.

In order to manage, record, and store the data gathered by the sensors, it is necessary to connect the chosen sensors to a system such as a building management system or a control panel. A Siemens Fire Safety System with the capability of

Fig. 2.2 Customized
Siemens fire safety
system—fire alarm control
panel

incorporating a security system was used to manage the data from the smoke detectors, contact sensors, and occupancy sensors. A Desigo™ Fire Safety System, 252–Point Fire Alarm Control Panel—Model FC 2025 is determined to be the optimal system. The customized system was installed into a portable weather-proof rolling case and is shown in Fig. 2.2. The fire alarm control panel is on the top right side of the case, and the security control panel is on the bottom of the right side, both are networked together. All of the recordings collected by the control panel are exported into a program TeraTerm. TeraTerm is a terminal that creates text documents that are records of the measurements made by the smoke detectors, contact sensors, and occupancy sensors over time.

A multi-criteria fire detector, Model FDOOTC441, was determined to be an optimal sensor as it can measure multiple types of data using optical, thermal and CO sensors. Smoke detectors are a prevalent reliable sensor used throughout most buildings, as they are required in the United States by most state and local laws as defined by NFPA 72 (National Fire Protection Association, National Fire Alarm and Signaling Code). Direct measurements of temperature and smoke obscuration as well as temperature, smoke, and CO alarms were relayed to the panel. The multi-criteria smoke detectors were located throughout the MFRI structural fire fighting building. The coordinates of the smoke detectors are shown in the Appendix. The origin is taken from the bottom left corner of the BIM as shown in Fig. 2.3. Each floor consists of two–five rooms and the detectors are placed to cover each of the rooms. The spacing of the detectors does not exceed the recom-

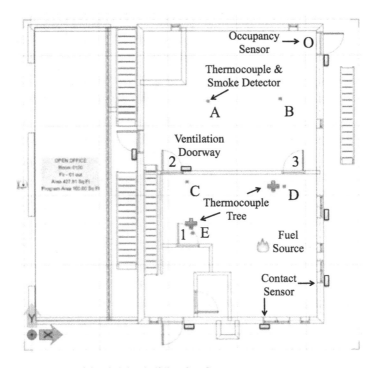

Fig. 2.3 MFRI structural fire fighting building first floor

mended nominal spacing of 30 ft of NFPA 72. This instrumentation layout gives high density of data that would be similar to an actual commercial building's instrumentation density.

Contact sensors measures the ventilation conditions by monitoring state of the windows and doors of whether they are closed or open. The contact sensors were installed on the first floor on all of the exterior windows and doors, along with the interior door that is closed or open for certain tests. The locations of the sensors are shown in Fig. 2.3. Contact sensors were chosen because they are prevalently used in modern infrastructure for security purposes. Price and Neviackas also determined that knowing the open or closed state of a door or window to a fire room is essential for calculating the fire size and location [14, 15]. Using contact sensors in a fire test will help determine if their recordings could be used for a cyber physical system that would aid fire safety.

Other than knowing the fire environment, one of the most important pieces of information for a fire fighter is the location of an occupant. Past literature and interviews with fire fighters all conclude this. A commonly used sensor in energy efficient buildings is an occupancy sensor. An occupancy sensor detects motion in a room and sends a signal to either turn lights on or off, depending if an occupant is in a room. Occupancy sensors were used in the fire experiments to observe their capabilities of detecting movement. Using occupancy sensors in the fire tests is a step in determining if these sensors are able to convey this information and can be used for further cyber physical systems development. The locations of the sensors are shown in Figs. 2.3, 2.4, 2.5, and 2.6.

Fig. 2.4 MFRI structural fire fighting building second floor

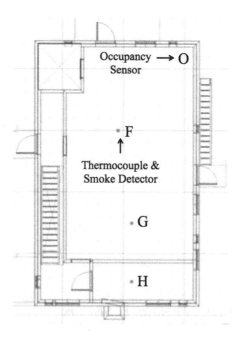

Fig. 2.5 MFRI structural fire fighting building third floor

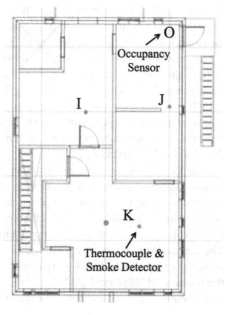

Fig. 2.6 MFRI structural
fire fighting building fourth
floor

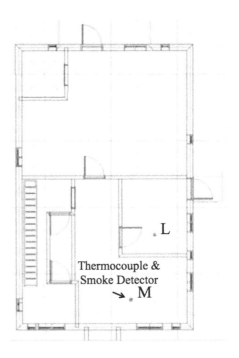

Thermocouple &
Smoke Detector

2.1.3   Laboratory Sensors

Laboratory sensors such as K-type thermocouples, a load cell, data acquisition
center, and a moisture content probe were used in this project to gather precise
measurements of the fire environment. This data will be compared to commercial
sensor measurements and used for future inverse modeling validation. Several
K-type thermocouples were used in the experiments to measure temperature
throughout the building. These K-type thermocouples have a maximum error of
0.75 % for readings above 0 °C. The measurements were taken every second
throughout the fire experiments. The K-type thermocouples were at all of the smoke
detector locations, and two trees of ten thermocouples were in the ventilation
doorways on the first floor.

The thermocouple locations are shown in Fig. 2.3. The thermocouples were
located adjacent to each smoke detector so that the measurements made at each
location can be compared to corresponding commercial sensor measurements
obtained from the smoke detector. The thermocouple trees each have ten thermocouples evenly spread out to give a temperature distribution from the ceiling to the floor.
The thermocouple tree, using a temporary stand set up, is shown in the appendix.
As in the experiments done by Price [14], the ventilation doorways were chosen for
the location of thermocouple trees. Using the temperature distribution and smoke
layer height, the average hot gas layer temperature could also be calculated for

inverse modeling interests [14]. Providing a thermocouple tree at each ventilation door also quantifies the ventilation status of the fire.

During the experiments, a load cell placed below the fuel package measured the weight of the fuel as it burns. The load cell measurements are accurate to 0.1 g and are taken at 1 Hz. The load cell measurements will provide the mass loss rate data of the fuel during the fire. This data will be useful for future inverse model development and describes the fire's burning characteristics. A heat shield was used to protect the load cell from the heat of the fire. The load cell was only used for the crib fuel load. The thermocouples and load cell are connected to a data acquisition center that was located outside of the MFRI structural fire fighting building. The data acquisition center was located a safe distance away from the building, was covered from the elements using a tent provided by MFRI, and was located next to the control panel. The data gathered was transferred and stored in the data acquisition center and was used for analysis.

2.1.4 Fuel Package Location

For training exercises at the facility of interest, fuels are typically located on the first, third, or fourth floors. For this project, the first floor was chosen as the ideal location for the tests. The first floor has a larger burn room, which limits the possibility of extreme conditions within the room. The location of the fuel package where the fire will take place is shown in Fig. 2.3. By placing the fire on the first floor, the smoke will rise and spread to the upper floors and throughout the entire building. The internal stairway is made of grated metal stairs, which will allow the smoke to move easily upward through the stairwell.

2.1.5 Fuel Package and Ventilation Conditions

The fuel and ventilation conditions used in testing were well characterized so that the resulting fire environment was controlled and the data gathered can be used for future inverse model validation. These conditions were guided by the practices used in typical training exercises at MFRI, by past studies, and by FDS simulations. The testing matrix defines two different ventilation conditions and two different fuel sizes. Various fuel types and ventilation conditions from past experiments and training exercises were simulated using FDS. The four cases ultimately chosen in the testing matrix result in controlled fire environments that are significantly unique from each other.

The size of the fire was optimized so that the smoke and temperatures could spread throughout the structure while still having a certain fire behavior in the fire room. The fire in the fire room could not be so large that would exhibit extreme temperatures for a long period of time which would result in the destruction of the

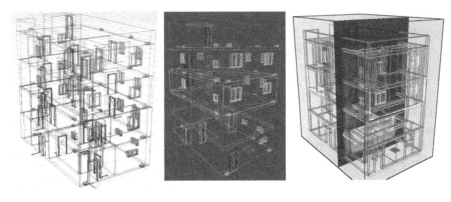

Fig. 2.7 Transformation of BIM File PyroSim to FDS File

building's fire resistant panels, or result in a ventilation limited fire. The fire was intended to burn out before conditions become too severe in the fire room and would therefore not have to be extinguished by the MFRI personnel. Since this environment was prescribed, the data can be collected throughout the entire lifetime of the fire. From the developing stage, to peak, to decay, and even ventilation data after extinguish can also be gathered. This will provide a significant contribution to future inverse model validation.

The FDS model used for the simulations was created from the BIM model. The BIM model was originally a MicroStation file. The MicroStation file to was saved as a CAD (Computer Aided Drawing) file. The CAD file was then imported into FDS using PyroSim. PyroSim, a program created by Thunderhead Engineering, is a graphical user interface for FDS. It allowed for the complexities such as building geometries to be transferred. The transformation between drawing files is shown in Fig. 2.7. Some building elements such as stairways and handrails were deleted to reduce to complexity of the FDS file. This was done to limit the computational power needed to run the simulations.

The two different ventilation conditions refer to the state of Door 2, which is located on the first floor. As shown in Fig. 2.3, Door 2 connects the fire room to the other room on the first floor. One case is that Door 2 is fully open and the other case is that the door is fully closed. All of the doors inside the building, other than the prescribed closed door in one case, were open to allow smoke to freely travel. All exterior windows and doors were closed. These two different cases result in different ventilation conditions for the fire room and affects the resulting fire environment. These two different scenarios were simulated in FDS to determine the resulting fire environment. Case 1, with Door 2 open, provides more ventilation to the fire. The resulting temperature slice file in the burn room at a time of 520 s is shown in Fig. 2.8. It can be seen that the highest temperature is around 170 °C and occurs within the 1 or 2 ft below ceiling height. Case 2, with Door 2 closed, reduces the ventilation area and the resulting fire environment at the same time of 520 s is shown in Fig. 2.9. The same maximum temperature of 170 C is exhibited but is now

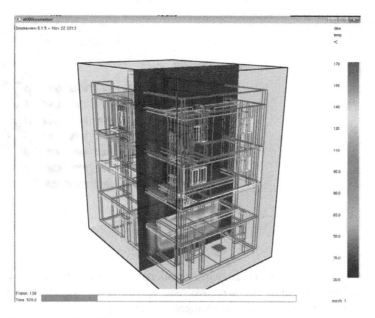

Fig. 2.8 Temperature Slice File for Case 1, Door 2 Open

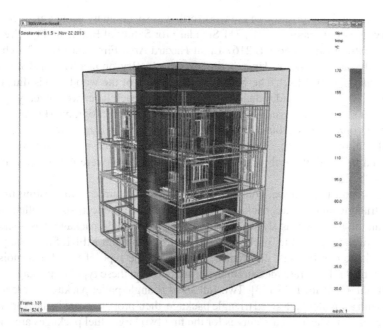

Fig. 2.9 Temperature Slice File for Case 1, Door 2 Closed

Fig. 2.10 Wood crib configuration

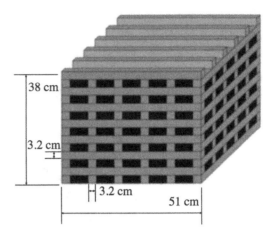

4–5 ft below ceiling height. These two simulations were similar but unique enough for comparison. The same fire size was used for these two simulations, similar results occur when the fuel size was changed respectively.

A wood crib was used as the small fire size, with an expected peak heat release rate of approximately 400 kW. This was calculated using a fuel burning rate of 22 g/s [19] and an effective enthalpy of combustion of wood of 19 kJ/g [20].

$$22 \ \text{g/s} * 19 \ \text{kJ/g} = 418 \ \text{kW}$$

The size and composition of the wood crib was specified by using guidance from Underwriters Laboratories (UL) 711 Standard for Safety of Rating and Fire Testing of Fire Extinguishers and UL 2167 Light Hazard Area Fire Test [21]. The chosen wood crib was the same used in experiments done by Bryson Jacobs at the University of Maryland in 2011 [22]. The general configuration of the wood crib is illustrated by Fig. 2.10 [22]. The individual wood members in each layer were evenly spaced forming a square. The average crib mass was around 21 kilograms (kg) and mass moisture content was around 5 %. The exact weights and moisture contents are provided in the Appendix. A controlled amount of excelsior, a fuel used by MFRI, was used to ignite the crib as petroleum based products were not allowed in the training facility.

The second fuel source, considered the large fuel source in the testing matrix, was a triangle shaped fuel package consisting of three reduced sized pallets and a half-bale of excelsior, as shown in Fig. 2.11. This triangle package was a smaller version of the typical fuel package used for fire training at MFRI. Since the pallet fuel package can be very different depending on the type of wood and moisture content, heat release rate measurements from NIST on these types of packages were used as the input for FDS [23]. Two full sized triangle pallet packages were tested along with a smaller half-sized triangle package, their properties are in the Appendix.

The resulting heat release curves for the first two larger fuel packages are shown in Fig. 2.12 [23]. The heat release rate of this smaller triangle is shown in Fig. 2.13 [23]. The expected heat release rates of the fuel were modeled in FDS to predict the

Fig. 2.11 Triangle pallet
configuration

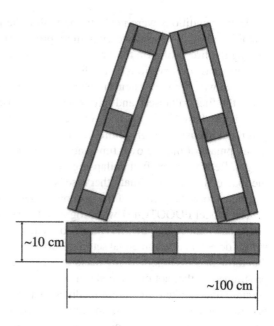

resulting environment as presented previously. It was determined using FDS that
full sized pallets triangles resulted in too large of a heat release rate for the compartment and the fire quickly became ventilation limited. The smaller pallet however,
resulted in a desired fuel limited scenario. The wood crib was also modeled in FDS
to observe the resulting environment and gave the desired fuel limited scenario. The
resulting fire environments for the crib and small triangle fuel configurations were
also determined to not reach temperatures that would threaten the building's fire
resistant panels.

2.1.6   Experimental Methods

The test matrix was designed using the sensors and fuel load selected; it was then
determined how to execute the experiments using the planned set up. Before each
test, the ambient outside temperature was noted, the mass of the fuel was weighed,
and the test start time was noted. These recordings and procedures are presented in
the Appendix. After the initial conditions are recorded, the final set up of the tests
was conducted. Experienced fire fighters set up the fuel as described in the previous
section. Photos of the set up are included in the Appendix. The inside doors are all
opened, and the outside doors and windows are closed. Once the set up was ready,
the team initiates device recording by starting the data acquisition center and control
panel. Within seconds, the fire fighters were notified via a handheld transceiver to
ignite the excelsior with a lighter.

As the conditions were not extreme within the fire room, the fire fighters could remain within the room wearing their protective gear and breathing apparatus. During the fire evolution the fire fighters were in communication with the team to convey visual observations of the fire. The visual observations consist of when the excelsior was ignited, when the excelsior ignited the wood crib or wood pallets, when the fuel collapsed, and any other visual observations. The event times were recorded and used in the analysis. One crib fire evolution and one pallet fire evolution was recorded using a high definition video camera. The videos provide insight to the development of the fire over time and various burning characteristics such as flame height. Recording the fuel collapsing also shows how much of the fuel falls off of the load cell, a quantity that otherwise would not be known. The fire fighters were instructed to walk in front of the occupancy sensors at certain times.

The Model FDOOTC441 was programed to convey smoke obscuration measurements and temperature measurements to the panel. This is not a typical capability of smoke detectors in commercial settings, but for this application we are able to achieve this through the support of engineers at Siemens. The detectors and the panel were programed so that not only the alarm state was communicated but also the raw temperature and smoke obscuration readings that cause the alarms. It was attempted to program the communication of the raw CO readings but the effort was unsuccessful. During the tests a team member pushed a query button to request temperature and smoke obscuration readings. A photo of this is shown in the Appendix.

2.2 Virtual Environment

An objective of this project was to transform the data collected in the full-scale experiments into information that would be useful for fire safety if these tests were to be an actual fire scene. Two types of data are collected; one is static data such as the building geometries and materials, and the other is dynamic data such as room temperatures and smoke obscuration levels. The data was visually represented in a virtual environment using a Building Information Model (BIM). Two Bentley products are used to display the static and dynamic data as fire safety information in this virtual environment: MicroStation information modeling software along with AECOsim Building Designer. The visual layout and quantity of the information is determined using guidance from past literature and caters to the capabilities of the Bentley software.

2.2.1 Program Selection

As a part of the SFPE Chief Donald Burns Memorial Grant, Bentley donated access to their various software programs. Many programs for building analysis, design, and information modeling are available. Two products were determined to be the most practical and useful for this particular project. The first is MicroStation; a 2D and 3D Computer Aided-Design (CAD) and information modeling software that can model, document, draft, and map projects of virtually any shape and size. It can produce lifelike renderings and animations and simulate the performance of the building. These capabilities make it an ideal program to utilize for simulating a life-like representation of an actual building fire environment. It has also become an increasingly prevalent tool for construction because it can also resolve utility clashes and simulate schedules. With current buildings be designed using this program, those files could also be utilized to carry out the future goal of this project of visually conveying a fire environment.

MicroStation information modeling software is the platform for architecture, engineering, construction, and operation of all infrastructure types including utility systems, roads, bridges, buildings, communication networks, etc. AECOsim Building Designer is determined to be the ideal program to use with MicroStation since this project virtually recreates the MFRI structural fire fighting building. AECOsim Building Designer allows a solid mass to be transformed into a collection of building elements such as walls, roofs, windows and doors. Intelligence is added to these objects by storing information in a collection of files called a dataset. The combination of MicroStation and AECOsim Building Designer provides the capability to recreate the MFRI Building with all of its building elements and the environment created by the full-scale tests.

2.2.2 Program Information Organization

In order to utilize MicroStation and AECOsim Building Designer to store and display the information about the MFRI structural fire fighting building, it is necessary to understand how the programs organize the information. The dataset information is organized into two main areas; the Family and Part System and the DataGroup System. Datasets are split to allow project specific information to be located at a project location, while maintaining an overall building dataset that stores information that can be shared by many projects. The Family and Part System defines a building object's graphical representation in both the 3-D model and in 2-D drawing views. The Part properties include the model definition, drawing symbology, cut patterns, centerline symbology, rendering properties, and possible structural and analytical information.

The original purpose of these products is for creating building information models of construction projects where engineers can virtually design their systems within the model. Design teams for every discipline need a system that enables them to assign important model data to objects to distinguish their use. Whereas the same Family and Part information can be applied to many objects in the building model, DataGroup information is individually applied to each object that stores DataGroup information. Assigned catalog item data is placed with each item instance and the system also tracks and manages this data for schedules and reporting (Figs. 2.12 and 2.13).

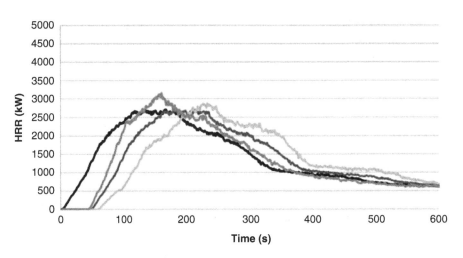

Fig. 2.12 HRR curves of triangular fuel packages [23]

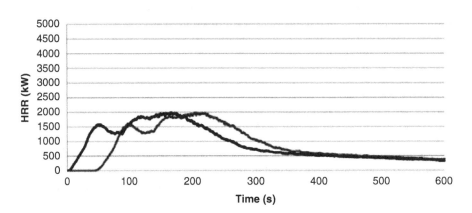

Fig. 2.13 HRR curve of small triangular fuel package [23]

2.2.3   MFRI Structural Fire Fighting Building Data Collection

In order to input the information about the MFRI structural fire fighting building into MicroStation and AECOsim Building Designer, the data had to be gathered through measurements and observations. Using the floor plans from the University of Maryland website, the MFRI structural fire fighting building was observed and measured with a MFRI employee. Each room was measured for its complete dimensions and recorded onto the scaled floor plans. The location and size of the doorways, windows, and stairs were all measured and photographed. Additional photos are included in the Appendix.

2.2.4   MFRI Structural Fire Fighting Building Data Transformation

The MFRI Structural Fire Fighting Building was modeled in AECOsim Building Designer V8i using MicroStation. A comparison of the photo of the building and the comprehensive BIM is shown in Fig. 2.14. The model has the exact dimensions as the measurements made within the actual building. Exact building materials were used for the walls, floors, doors, windows, stairs, and other building elements. Only the fire resistant paneling could not be modeled as there was not an option for this specific type of wall. The exterior walls are made of 1 ft thick brick, the floors are 1 ft concrete slabs, and the interior walls are either 6 in. CMU (Concrete Masonry Unit) or 4 in. thick brick walls. The windows are single or double metal casement. The doors are metal with metal frames. The scuttles are metal louvers with metal frames. The stairs are made of metal grates and metal piped railings. There three access panels on the roof that can be used for ventilation that are shown in the model. Building history information like construction date, past and current uses, number of floors, square footage, and building construction were also included in the model. Each room was modeled as a specific space and was automatically associated with a room number, floor number, and square footage.

The smoke detectors were imported into BIM using a Revit file from BIM file database that contains many different products. The file contains inputs for many properties of the detector such as cost, assembly code, and manufacturer; these properties are shown in the Appendix. When a detector is placed within a room, it is associated with that space. The exact coordinates of the smoke detectors within the BIM are the same the locations of the detectors in the fire tests.

Fig. 2.14 Comparison of building photograph and comprehensive BIM

2.2.5 Static and Dynamic Data Transformation to Information

A tremendous amount of data can be stored within a BIM. Not all of that data will be useful to show to a fire fighter during an event. The data rather could be used for fire modeling calculations to determine information that is worth showing such as the fire size and location. Important types of information to know when assessing a fire environment are room geometries, wall thermophysical properties, ventilation conditions, and fuel load. These are the basic inputs for a fire dynamics analysis or simulation. These initial conditions combined with and the measured fire environment characteristics such as temperature, smoke obscuration, and CO concentration can be used for an inverse fire model to determine the fire size and location. This project defines how this information can be utilized from a BIM and from commercial sensors.

Many of the inputs for a fire dynamic analysis are already imbedded into a BIM model when it is designed. For example, when each room is created a floor number and square footage is automatically assigned to it based on the geometry. In Fig. 2.3 the room description is left on for the room on the left to show the automatic assignment of the square footage. When a wall is created in BIM there are many options for the designer to choose from such as brick, CMU (Concrete Masonry Units), or gypsum drywall that have embedded properties such as fire resistance ratings, and surface spread of flame values. These values are determined and referenced from the appropriate UL standard, and NFPA 255, Standard Methods of Test Surface Burning Characteristics or ASTM E 84. These values can provide insight into the proper thermophysical properties that could be used for them within a fire dynamics analysis. In the approach, the BIM created is imported into FDS and automatically provided all of the geometry information and thermophysical properties of the walls. This project takes this process a step further by setting up how more information stored within a BIM can be used for fire modeling.

Room geometries, wall properties, and fuel load information are all static information components that can be stored within a BIM. To complete the analysis, dynamic information about the current fire environment can to be transferred from sensors into a inverse fire model and combined with the static information to determine the fire size and location. Although this is process is not determined within this project, the data collected during the full-scale fire tests could be used to carry out this process. The sensors are chosen to provide the necessary information needed to conduct an inverse modeling analysis. The measurements obtained are temperature, smoke obscuration, carbon monoxide, and open or closed condition states of doors and windows.

Each element can store specific data that is unique to that device. An imported Siemens smoke detector file can be manipulated to create new properties associated with the device. Properties that were assigned to the specific detector within this BIM are coverage area, mounting height, humidity, smoke obscuration, and temperature, as shown in a figure in the Appendix If these values could to be linked to actual sensors readings in the built environment it would provide the temperature within an environment and the location that the temperature was recorded at. Ventilation conditions are another aspect that can determine fire information. The window and door geometries and locations within BIM combined with contact sensor information can provide the ventilation conditions for a fire.

A recent study conducted by Anderson developed a framework to provide an informed risk analysis of a building using a combination of BIM software, fire models, and statistical analysis. Fuel packages are defined in this framework based on a statistical analysis. While this process was intended to be used with forward modeling techniques, its method can be utilized and combined with the inverse model framework developed for this project. Fuel load information can provide insight into the fire size, fire location, and possibility of fire spread.

Other aspects that could be included in the model for fire protection purposes are egress components and locations, location of roof access, hazardous materials, and location and description of fire hydrants. Also fire department hookups for sprinkler system/standpipes, staging areas with entrances and exits to building, location and description of fire alarm panel and remote annunciator panels, and areas (zone boundaries) protected by sprinklers or other devices are all important considerations to include in the model since they provide useful information to fire safety officials.

Chapter 3
Results

The cyber physical system test bed created for fire safety provides insight for its feasibility for future fire fighting strategies. The data collected during the fire tests can be used for future inverse model development and validation. The visualization exploration examines current fire safety needs and BIM capabilities. This cyber physical system test bed framework is tested using the approach outlined in the previous section. The observations and results of the fire tests conducted are presented in this section. Data from fire tests in the MFRI Structural Fire Fighting Building using laboratory sensors and Siemens commercial sensors are evaluated using simple fire dynamic principles. Using guidance from previous research the fire state is visualized in BIM and through a timeline format presented in this section.

Each fire scenario defined in the test matrix in Table 2.1 was executed at least twice, and a total of ten tests were conducted. The initial conditions such as ambient temperature, fuel mass, fuel moisture content, and start time are included in the Appendix. Test event times such as fuel ignition, occupancy sensing, fuel collapse, and ventilation changes are also provided in the Appendix. Examples of the raw data collected by the Siemens system of temperature, smoke obscuration, and security switches are in the Appendix.

The four cases of the test matrix as shown in Table 2.1 consisted of different two fuel loads, and two different ventilation states. Case one was the crib fuel load with the ventilation doorway, door number two in Fig. 2.3, as open. Case two was the crib fuel load with door two closed. Case three and four are the pallet fuel load with door two or closed. The most interesting case of one crib fire and one pallet fire test, case two and case four, were chosen and their results are provided in this section.

© The Author(s) 2016
R.F. Wills, A. Marshall, *Development of a Cyber Physical System for Fire
Safety*, SpringerBriefs in Fire, DOI 10.1007/978-3-319-47124-2_3

3.1 Timeline

A timeline was created to provide an overview of the events that occurred during the test and when they occurred. While real-time fire information visualization is important, the history of a fire event can also provide important information about the fire. Both the timing of the events, as well as what events have occurred in the past can provide insight into understanding the fire. For example, knowing which smoke detectors went into alarm at what times can provide insight into where the fire started and how quickly the fire developed. Both reports written by NIST identified two different types of information presentation, one to be shown enroute and one for onsite [17, 18]. NIST identified the enroute presentation to be about the area surrounding the building, where the incident was located and the presence of unusual hazards [18]. For this project, a timeline is developed to be used as the main source of information for the enroute stage of an event, and could also be used on site. The timeline developed for this project was manually created, it is envisioned that a timeline would be automatically created in the future through communications between sensors, a building management system and BIM.

The events chosen for the timeline are those that occur at a single time such as a smoke detector going into an alarm state, or the fuel collapsing. The events are only the measurements taken by the commercial sensors, events associated with the commercial sensors, or events of the fire that are assumed because of the measurements of the commercial sensors. The events are best represented by a timeline as each event's location and relation to other events provide insight into the fire development and how the sensors responded to the environment over time. The timeline for a crib fire, the smaller fuel load, is presented in Fig. 3.1. The timeline for a pallet fire, the larger fuel load, is presented in Fig. 3.2. The timeline provides a visual reference for the history of the sequence of events throughout the entire fire duration.

3.1.1 Crib Timeline (Small Fire)

Once the excelsior ignited the crib after 1 min, the crib burned at a steady state for about 12 min. During that time, temperatures (Significant Readings) on the first floor, in the burn room and in the adjacent room increased and remained steady until the crib fire started to decay. The smoke traveled to the upper floors and throughout the building. During this time, the smoke detectors responded (Local and Remote Alarms) with smoke alarms, temperature alarms, and CO alarms, the times are shown in Figs. 3.1 and 3.3. About 12 min into the fire, the crib collapses (Fire Event) from its original configuration. Immediately after, the smoke detectors report higher temperatures for a period of a few minutes. After 20 min the ventilation of the fire floor is initiated and the contact sensors report their open status to the panel (Security Readings). During the ventilation stage the temperature and smoke obscuration measurements reduce.

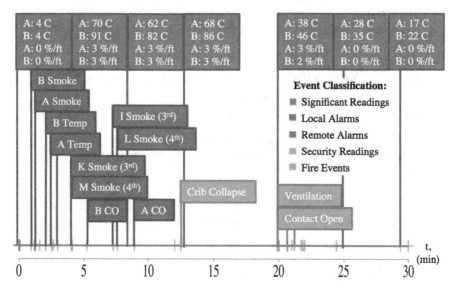

Fig. 3.1 Timeline of crib test (small fuel)

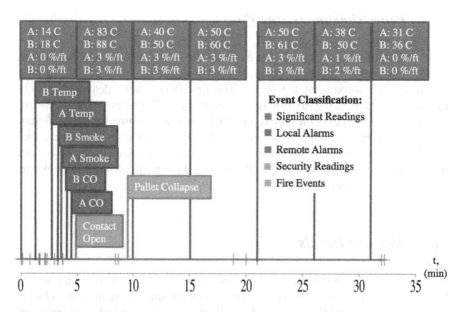

Fig. 3.2 Timeline of pallet test (large fuel)

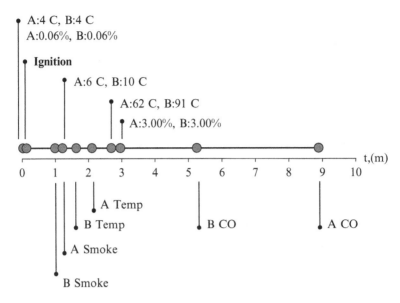

Fig. 3.3 Beginning timeline of crib test (small fuel)

3.1.2   Pallet Timeline (Large Fire)

For the fire few minutes the initial fire environment is governed by the large amount of excelsior used in the fuel. The large amount of smoke produced by the excelsior quickly spreads throughout the floors. The first floor smoke detectors quickly respond with temperature alarms, smoke alarms, and CO alarms (Local Alarm). Once the excelsior decays the pallets provide a sustained burning for about 10 min. Similar to the crib fire, temperature increase was noted just after the time the fuel was observed to collapse (Significant Readings). For this particular test, the contact sensors reported an open state at an early stage of the fire and no upper floor alarms were initiated (Security Readings).

3.1.3   Timeline Details

The ambient conditions, measured by smoke detectors A and B in room one, are described in the first event shown in blue. Significant smoke obscuration and temperature measurements for detectors A and B throughout the rest of the experiment are also provided on the timeline (Significant Readings). These measurements are events shown in blue. Detectors A and B are the closest detectors to the fire. They are located in the room adjacent to the fire room, which provides important

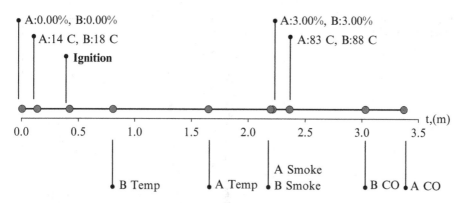

Fig. 3.4 Beginning timeline of pallet test (large fuel)

quantitative measurements to describe the fire environment and can put the events of the timeline into perspective. Other unique events are written out: "crib collapse" or "pallet collapse" is when the fuel falls apart and looses its original configuration (Fire Event). This event is included as the smoke detectors reported the temperature elevated after the temperature were decreasing, providing insight that the fuel had changed. "Contact Open" is when a contact sensor reports an open state to the panel (Security Reading). "Ventilation" is when the windows to the outside are opened on the first floor, and was known from the contact open signal (Fire Event). The events shown in red are when detectors A and B go into an alarm state: "smoke" is for a smoke alarm, "temp" is for a temperature alarm, and "CO" is for a carbon monoxide alarm (Local Alarm). A closer look at the exact times the alarms are initiated for the crib test and pallet test are shown in Figs. 3.3 and 3.4.

3.2   Test Measurements

3.2.1   Siemens Smoke Obscuration

The recorded smoke obscuration levels for the crib test and pallet test are displayed in Figs. 3.5 and 3.6 respectively. The four different colors correspond to the four different floor levels: the first floor is shown in red, second floor is purple, third floor is blue, and the fourth floor is green. As discussed in the approach, the reported smoke obscuration is the amount of smoke needed to go into alarm, not the amount of smoke in the chamber. Therefore, using the 3 %/ft threshold, the raw smoke obscuration readings are converted into the actual smoke obscuration levels within the fire environment.

Fig. 3.5 Smoke obscuration of crib test (small fuel)

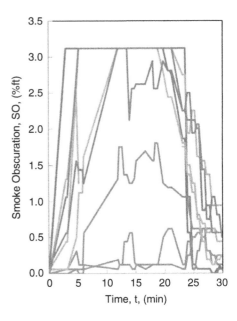

Fig. 3.6 Smoke obscuration of pallet test (large fuel)

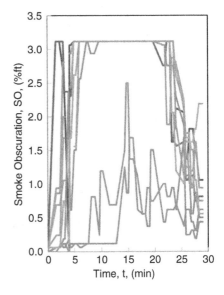

3.2.2 Siemens Temperature Readings

The smoke detector temperature readings for a crib test and for a pallet test are shown in Figs. 3.7 and 3.8 respectively. The same colors are used for the different floors. Detector A and B are distinguished by dash length, temperature measured by detector A is shown with long dashed lines and short dashed lines for

Fig. 3.7 Siemens
temperature of crib test
(small fuel)

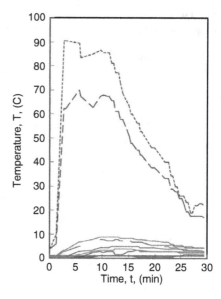

Fig. 3.8 Siemens
temperature of pallet test
(large fuel)

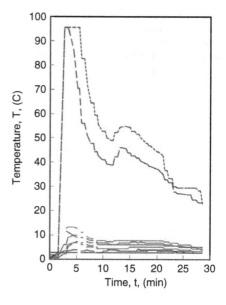

detector B. The gaps within the upper floor data are when the system is unable to
communicate the temperature readings for the fire because it is interrupted by
alarm warnings.

Raw thermocouples readings for each detector location are compared and the
most interesting cases are presented here. Detectors A and B are concluded to be the
most interesting cases. The temperature measured by the detectors is compared to the

Fig. 3.9 Siemens
temperature and
thermocouple temperature
at locations A and B of
crib test (small fuel)

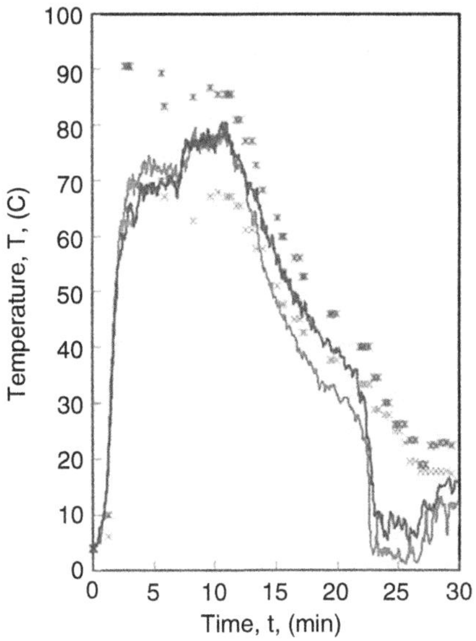

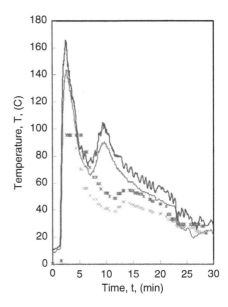

Fig. 3.10 Siemens
temperature and
thermocouple temperature
at locations A and B of
pallet test (large fuel)

temperature measured by the thermocouples for one crib test and one pallet test in
Figs. 3.9 and 3.10 respectively. Location A for both the thermocouple and the detector is shown in blue, and red for location B. The smoke detector readings are the "X"
locations, and the solid lines are the thermocouple measurements. These figures provide insight into the accuracy of the sensors as compared to the thermocouple measurements, as well has how frequent the data from the detectors can be conveyed.

3.2.3 Thermocouple Temperature

The raw data for all of the temperature measurements for one crib test and one pallet test is presented in Figs. 3.11 and 3.12. Only the top measurement from the two thermocouple trees are used for these figures. This raw data shows the temperatures throughout the first, second, third, and fourth floors. As before the four different colors correspond to the four different floor levels: the first floor is shown in red,

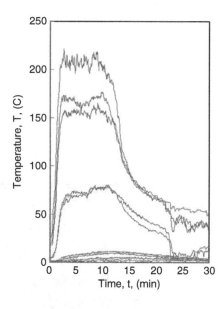

Fig. 3.11 Raw thermocouple temperature of crib test (small fuel)

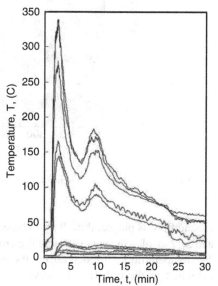

Fig. 3.12 Raw thermocouple temperature of pallet test (large fuel)

Fig. 3.13 Raw
thermocouple temperature
of two crib tests; location
A (Red); B (Blue); C
(Green); D (Purple); E
(Orange); Trial 1(Light);
Trial 2 (Dark)

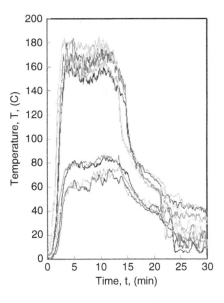

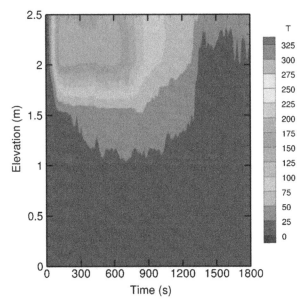

Fig. 3.14 Thermocouple tree temperature of location D of crib test

second floor is purple, third floor is blue, and the fourth floor is green. Figures 3.11
and 3.12 display the very different fire environments created by the two fuel sources.
The temperatures exhibited during the pallet test are significantly larger than those
exhibited in the crib test. However, the temperatures are sustained for a longer

period of time during the crib test. In Fig. 3.12 the temperature significantly increases right after when the pallet triangle fuel package collapses.

To provide insight into the repeatability of the experiments, the raw temperature measurements for the first floor for Test 1, Trial 1 and Test 1, Trial 2 are shown in Fig. 3.13. The darker colors are for Trial 1 and the lighter colors are for Trial 2.

The colors correspond to the location of the thermocouple, location A is red, location B is blue, location C is green, location D is purple, and location E is orange. For locations D and E the third highest thermocouple was chosen. It can be seen in Fig. 3.13 that the measured temperatures are similar for the two trials.

Contour plots for each of the two thermocouple trees over time is presented in Figs. 3.14, 3.15, 3.16, and 3.17. The contour plots provide a visual for the temperature distribution along the height of the compartment throughout the entire fire evolution. The raw 1800 point data is averaged for every 10 s interval to provide the 180 inputs for the contour plot. The contour plots provide insight into the ventilation state of the fire.

Recently, Price conducted multi-room compartment fire experiments to obtain measurements of hot gas layer temperature and depth [14]. These measurements were conducted by using thermocouple trees at the ventilation doorways just like the thermocouple trees used in this study. These measurements are used by Price to determine the location and size of a fire using an Inverse Fire Model (IFM). The measurements taken in this experiment can be used for future IFM validation. Once an accurate process for using an IFM is developed, fire size and location could be conveyed to fire fighters, which is one of the most significant elements of information during a fire event.

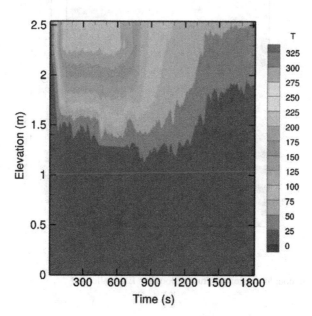

Fig. 3.15 Thermocouple tree temperature of location E of crib test

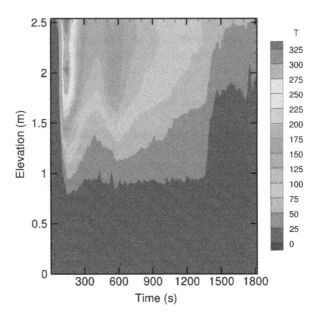

Fig. 3.16 Thermocouple tree temperature of location D of pallet test

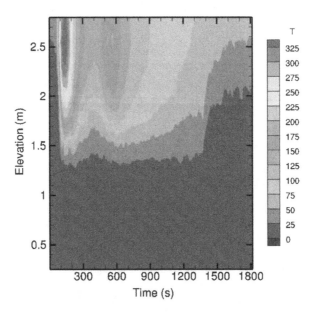

Fig. 3.17 Thermocouple tree temperature of location E of pallet test

Fig. 3.18 Mass burning
rate of crib test (small fuel)

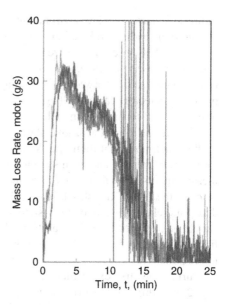

3.2.4   Mass Loss Rate

Three tests successfully measured the mass of the crib throughout the duration of
the fire evolution. The three results are presented in Fig. 3.18. The raw data is averaged between three points to smooth the data. A five point derivative average is then
applied to the smoothed average to get the resulting mass loss rate (mdot) presented
in Fig. 3.18. The five point derivative method is taken from Eq. (9) of Section
13.3.2 in ASTM E1354 [24]. The mass loss rate for each crib is similar for the first
10 min of the experiment. The prescribed crib configuration provided controlled
steady burning. The cribs collapsed between times of 12–14 min. The mass loss
measurements have a lot more noise during this time interval and could be caused
by the significant mass level change from collapsing.

3.2.5   System Limitations

During the first few minutes of each test, the system is overloaded with sending
alarm signals that the temperature and smoke obscuration measurements are not
conveyed. This is an unusual feature of the system that will significantly limit the
critical data needed to determine information about the fire. The first few minutes
and the developing stage of the fire was a very important time to be making temperature and smoke obscuration measurements. This needs to be repaired in the

future so that accurate fire environment measurements can be made. The only smoke obscuration readings relayed are the beginning measurements of 0.06 % and the maximum 3 % obscuration given 3 min later. After a detector reaches the 3.00 % alarm threshold, it is not capable of describing the smoke environment further. This is important to consider when determining if smoke obscuration measurements read by smoke detectors can actually give an accurate description of the fire environment or can be used for inverse modeling purposes.

Although some of the smoke alarms on the third and fourth floors were relayed to the panel, not all of them did. When the 3 % threshold is reached the detectors should have relayed an alarm state to the panel. The measured smoke obscuration measurements shown in Figs. 3.5 and 3.6 displays that most of the detectors reached the 3 %/ft threshold, but as shown in Figs. 3.1 and 3.2, not all of these detectors relayed a smoke alarm state to the panel. This is a system limitation and should be explored further. In all tests, there were no indications that when the fire fighter walked in front of the sensor that the detection of an occupant was relayed to the panel.

During the first experiment the smoke detector in location E, as shown in Fig. 2.3, within the burn room was ignited and a short circuit was reported. The single short circuit caused the entire Siemens device loop to go down during this experiment. The system still had power and the display was still up but the system was only displaying trouble messages. It was not expected for one device failure to cause an entire system to stop recording entirely. Later, it was realized that all of the devices were looped together, where if one fails, the entire system would fail. In real buildings, a few sensors may be looped together in certain areas, but codes and regulations limit this, as many sensors need to stay operating in the case of one failing. Disconnecting the other sensors within the burn room and connecting the sensors in room one onto a separate loop rectified this issue. In this case if A or B failed, the entire system would not go down. This solution worked but in future tests it is recommended to not disconnect the detectors in the fire room but rather place them on a separate loop. The other two detectors within the fire room were never ignited throughout all the tests and would have been able to stay operational.

3.2.6 Inverse Fire Modeling Validation Data

Some of the data collected was not intended to be incorporated into the BIM, as it is not something that is typically within a commercial building. Rather these measurements were taken to provide more rich dataset about the specific fires that occurred during the full-scale fire tests. Thermocouple tree data provides insight into the ventilation status of the fire, and the load cell provides insight into the mass burning rate of the fuel. These measurements can be used to determine the heat release rate of the fire.

3.3   Real-Time Cyber Physical System Framework

MicroStation information modeling software along with AECOsim Building Designer were used to create the virtual framework for the cyber physical test bed for fire safety. In this section the static and dynamic data incorporated into BIM is presented. The visual layout and quantity of the information is determined using guidance from past fire safety literature and caters to the capabilities of the Bentley software. A method to utilize inverse modeling techniques to describe the fire environment is presented.

3.3.1   Real-Time Critical Fire Information Visualization

A framework for visualizing a virtual a fire environment was developed. The framework utilizes a BIM and measurements collected from sensors during an actual fire scenario. The framework defines how the data is displayed based on past fire safety research. The previous section defined how the data can be transformed into information such as fire size and fire location. This section defines what dynamic information can be transferred directly into a BIM as the value itself is useful for a fire fighter and not just for a fire analysis that determined the fire size and location. The framework is carefully determined using the past fire safety research defined in the literature review.

There are two different stages of fire response, enroute and on the fire scene. The timeline could be used for both stages to provide a history of the fire development. A visualization of the current real-time information of the fire environment will allow EFRs to make informed decisions on the fire scene. An overview of the BIM created of the MFRI structural fire fighting building is shown in Fig. 3.19.

A visual of the entire building infrastructure will allow EFRs to quickly see the location of the stairs, exits, balconies, and if the structure had it; fire hydrants, standpipes, sprinklers, and HVAC utilities. It is important to realize that many of the building elements that were created as a part of the building design also provide information about the fire environment. A 3-D representation of the building allows for the fire fighter to know the floor layout beforehand. Navigating stairwells, exits, finding standpipes, are all important building information that can automatically be shown without real-time sensor information.

The view in Fig. 3.19 makes it difficult to see the complexities of each floor and therefore a single floor option is shown in Fig. 3.20. The floor of importance can be seen immediately in Fig. 3.19 through the use of color to show a certain temperature reached on that floor.

Another method of viewing the building materials is shown in Fig. 3.20. A wireframe method is shown in Fig. 3.19 and allows for the building elements such as the stairwells to be highlighted easily. This illustration method shown in Fig. 3.20 allows for the walls, doors, windows and overall geometry of the floor to be seen easily, just as it would be seen in actuality.

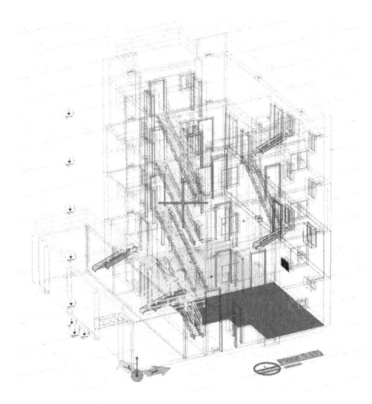

Fig. 3.19 Visualization of building and fire environment

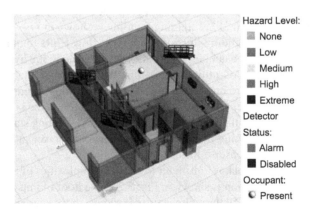

Fig. 3.20 Visualization of test bed fire environment

One snapshot of the fire environment observed during testing is shown in Figs. 3.19 and 3.20. The dynamic fire information is shown through methods determined from past studies and the capabilities of the BIM software. Room temperature is a significant measurement that defines the hazard within a specific area. Fire fighters and occupants can be exposed to certain temperature ranges for a limited amount of time. For this framework, these ranges are determined to have a certain hazard value and are associated with a certain color. "A Guide to Smoke Control in the 2006 IBC" is referenced by SFPE and the ICC as indicating that humans can typically tolerate continuous exposure to humid are at 66 °C for a period of 20 min [25]. Also, the first temperature warning threshold used by the NIST study is 50 °C and requires a breathing apparatus for fire fighters [18]. Madrzykowksi observed various fire training scenarios and determined that temperatures in excess of 260 °C and with heat fluxes in excess of 20 kW/m^2 would result in a survival time of less than 30 s for someone that was in full personal protective equipment (PPE) [23]. With this insight is it determined to provide four temperature thresholds with corresponding colors and hazard levels. The first warning is for when temperatures reach 50 °C, and the corresponding color is green. The second warning is for when temperatures reach 100 °C, and the corresponding color is yellow. The third warning is for when temperatures reach 150 °C, and the corresponding color is red. The fourth warning is for when temperatures reach 250 °C, and the corresponding color is black. The colors are assigned to each room and are placed on the floor of the BIM, as seen in Fig. 3.20. The solid colors on the bottom of the compartment allow the rest of the compartment to be seen, there will be no color assigned to it if there is no alarm.

It is determined that only an alarm state will be shown for the detectors since they can only measure up to the alarm threshold of 3 %/ft obscuration. A detector shown in red will signal an alarm state. If a smoke detector fails, the detector will be shown as black. Detectors in alarm and detectors that have failed are shown in Fig. 3.20. A sphere, as shown in Fig. 3.20, represents an occupant. This representation was also used in the NIST example discussed in the literature review [17]. The location of an occupant, taken from an occupancy sensor, would allow a fire fighter to recover a person in need safely. The ventilation status of the windows, transmitted from a contact sensor, is shown by either having an open clear window or a closed blacked out window as shown in Fig. 3.20. The visual cues described in this section that are incorporated into BIM to describe the fire environment had to be manually created. In the future this process would be automated using a macros and visual basic program. This project only focused representing the data that was gathered by the commercial sensors such as temperature, smoke obscuration, open or closed state of the windows, and occupancy detection. There are many more elements that can be visualized using BIM as described in the literature review, such as fire size and fire location.

3.4 Conclusions

A robust large-scale CPS test bed was formulated and assembled, producing a CPS dataset to explore the viability for supporting critical fire safety decision-making. Novel tests methods were developed for the CPS, which included careful selection of fire and non-fire sensors to observe controlled fire environments. Various realistic fire environments were designed by prescribing well-characterized fuel sources. The fuel sources varied in size and growth, with 1.5 MW peak HRR and 0.4 MW peak HRR. The executed test methods successfully recreate the fire environments simulated using FDS.

The physical infrastructure, which consisted of a multi-story building integrated with a customized Siemens fire panel collecting signals from conventional fire and non fire sensors and laboratory instruments. The sensor output was directed to a virtual framework designed for visualization of critical fire safety information. The virtual framework consisted of a detailed BIM representation and custom visual cues based on fire safety research guidance. Rich sensor data of a well-controlled fire environment provides valuable information for developing inverse fire models. Measured temperature profiles, smoke obscuration, ventilation areas, and mass burning rates can be utilized for inverse fire model validation. The CPS test bed developed produced remarkable evidence about the opportunities created by the communication between sensors, BIM, and fire for fire safety. The CPS test bed developed has many opportunities for expansion.

3.5 Future Work

A cyber physical system could provide infinite benefits to improve fire safety, as well as there are an infinite number of ways to develop that cyber physical system. This study aimed to focus on what key aspects of a fire environment are critical to know when determining an effective fire safety strategy. This specific formulated test bed, determined by the methods described above, could be utilized again many times. The test was created with repeatability in mind. Future students could easily reproduce this test bed using the methods described in this thesis. All resources utilized for this project, the MFRI training structure, the Siemens control panel, the Bentley Modeling software are all a part of a network that is interested in furthering cyber physical systems for fire safety. These tools can be recycled to replicate as much or as little of the test bed as possible. Further research could include introducing new variables, or providing additional investigation on the current variables that were considered in this project.

The Siemens panel was designed in a transportable weather-resistant shell. This allows for future testing to be completed easily at the previously chosen structure, or in virtually any other testing structure. Testing in different geometries will provide more insight into how compartmentalization will affect both the fire environment, as well as how it will affect the readings from the sensors that are detecting that fire environment. The panel can also operate with many other fire alarm and security devices. If another device is determined to be useful in measuring a key fire environment element, and is prevalently used in modern infrastructure, then it can straightforwardly be tied into the existing Siemens system. One possible option would be for fire alarm devices that are tied to the sprinkler systems such as valve monitors and tamper switches. This could provide information about whether or not a sprinkler has activated and in what area it has activated.

The Bentley program provides an unlimited number of possibilities for the type of structure that can be modeled within the software. As all buildings are unique with their own hazards and utilities, providing more information about each building could aid the fire fighters in their response. Providing the location of the fire command center, fire department connections, fire pump room, standpipes, fire hose valve connections, etc. would guide the fire fighters where to go in order to complete their tactics. It would be a huge benefit to push the boundaries of the software to visually represent a fire environment. Determining how to visualize alarmed devices and their data in a way that is valuable and concise would be useful. One development that would be useful for turning this vision into a reality would involve determining how a simplified model from Bentley software could be interfaced with a Siemens fire alarm control panel.

Also, the intent of this research was to provide data to validate inverse modeling. The raw data that was collected by the experimental and commercial sensors provide a rich description of the fire environment. This rich dataset will be the starting point for determining the fire size using inverse modeling. As we have already well characterized the fuel's heat release rate curve, the output of the modeling can be compared in detail. The data of the commercial sensors can be further analyzed against the precise experimental sensors and provide insight into how they characterize the fire environment. Limitations and margins of error could be assessed of the commercial sensors to determine how inverse modeling could ultimately be used with their data.

With technology rapidly advancing, some of these strategies may be able to be put in place soon. However, the overall concept of cyber physical systems for fire safety will be an evolving production that over time will change with the modern built environment and the virtual tools of that time. It is impossible to predict the various directions that each will take, but steps can be done now to figure out how these systems can be developed and implemented in order to understand how they could be used to improve fire safety.

Appendix

A.1 Facility Description

Building utilities such as HVAC and plumbing are not used within the building because there would be little use for them and would be damaged during the fire fighting training exercises. The walls are constructed of concrete masonry units or of brick, and the floors are made of concrete. The walls and ceilings are protected with different types of additional fire resistive materials. The first is a sprayed-on fire resistive material known as Pre-Krete G-8. Pre-Krete G-8 is composed of hydraulic calcium aluminate cement. The second type of fire resistive material used in the burn structure is 51 mm thick, high-temperature tiles composed of refractory concrete placed on top of a 25 mm thick insulation known as SuperTemp_L. The high temperature tile insulation combination is attached to the ceiling of the third floor. The third type of fire resistive material is Duraliner HT insulating panels. The first floor of the building recently was renovated with new Padgenite panels. Padgenite boards are made of calcium silicate and provide a thermal barrier for the walls on the first floor. The rest of the building elements are made of metal, which can withstand extreme conditions. The metal doors and windows are free swinging with a manual latch. The stairs inside of the building and the fire escape on the outside are grated metal stairs with metal handrails. Although these materials will interact with the fire environment differently than typical materials used in commercial buildings, the protected environment allowed for full scale burns to take place without risking the integrity of the structure.

The building is a part of the University of Maryland campus and is building number 196 of the building inventory. Within the facilities website there are floor plans and general details about the building. According to the website, the MFRI structural fire fighting building has a gross square footage of 5701 ft and 4828 net assignable square feet. It also states that the year of construction was in 1989, undergoes normal maintenance, and its function is non-academic. The replacement value is $3,082,117 and the renovation cost is $1,541,059.

© The Author(s) 2016
R.F. Wills, A. Marshall, *Development of a Cyber Physical System for Fire Safety*, SpringerBriefs in Fire, DOI 10.1007/978-3-319-47124-2

A.2   Siemens Fire Safety System Descriptions

The system can monitor 252 addressable devices, which is the smallest option that exceeds our need of monitoring 25 of addressable devices. The control panel has an LED screen display that provides detailed information about the nature and location of the event. During testing this is used to get an update on the condition of the fire tests. The control panel is connected to a security panel so that the data collected by the contact and occupancy sensors is recorded.

All smoke detectors are wired back to the fire alarm control panel. The hard wiring is necessary to relay the significant amount of data gathered by the detectors to the fire alarm control panel. The wiring used is Honeywell Genesis 41111004 16/2 Solid Unshielded Cable. The wiring can withstand high temperatures but cannot be in direct flame contact. The security sensors are wireless and are monitored by Siemens HTRI-D devices. The HTRI-D devices are dual-impute modules that are designed to supervise and monitor two sets of dry contacts and report the status to the control panel. These are included in the left side of the case, as shown in Fig. 2.2.

This system is used to gather data during fire tests at the MFRI structural fire fighting building. The system is designed be portable so that temporary installation can be achieved to not harm the integrity of the training facility. The weight, size, and power requirements of the two panels allowed for the system to be portable (Figs. A.1, A.2, and A.3).

Temporary installation techniques needed to be utilized as the facility conducts training exercises regularly. The system prescribed would not withstand the conditions of the training exercises if it were to be permanently installed. The system also could not be physically attached to the facility in a way that could damage the

Fig. A.1. New padgenite insulating panels on first floor

Fig. A.2. Metal free
swinging window with
manual latch

Fig. A.3. Metal grate
stairs with metal handrails

integrity of the fire resistant panels. Using adhesives and other non-penetrating attachment methods were determined to not be robust enough to stand up to the fire environment. The resulting installation design used for the ceiling mounted detectors is a temporary stand as shown in Fig. A.4. The stand is constructed of a metal Christmas tree stand and a metal conduit pole. The heights of the poles are determined based on the different ceiling heights. The smoke detectors are propped up against

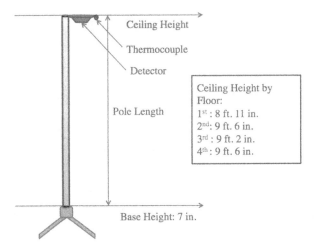

Fig. A.4. Temporary sensor instrumentation

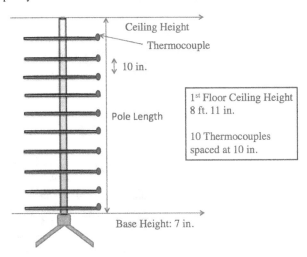

Fig. A.5. Temporary thermocouple tree instrumentation

the ceiling using a metal clamp that is attached to the top of the metal stand. The measurements at the smoke detectors are uniform at the highest elevation in each room. The highest elevation measurement was also chosen due to past studies observing that ceiling temperature can be related to fire size and is therefore an important measurement for inverse modeling. As shown in the figure, there was also a thermocouple provided at each smoke detector location and at the same elevation. To provide a more rich dataset closer to the fire, there are three smoke detectors in the fire room and two in the adjacent room on the first floor. Also in the fire room were two thermocouple trees at two of the doorways, the thermocouple tree arrangement is shown in Fig. A.5. Again, this was proven to provide signification data to interpret the fire environment for inverse modeling (Tables A.1, A.2, A.3, and A.4).

Table A.1. Sensor
coordinates

	X (m)	Y (m)	Z (m)
Sensor			
A	3.10	9.73	2.72
B	6.30	9.83	2.72
C	2.18	6.30	2.72
D	5.72	6.50	2.72
E	2.39	3.96	2.72
F	4.17	8.66	5.90
G	4.50	4.09	5.90
H	5.44	1.01	5.90
I	7.32	8.53	8.80
J	3.48	8.64	8.80
K	5.72	3.00	8.80
L	6.63	4.34	11.90
M	5.31	1.37	11.90
Thermocouple			
A	3.10	9.73	2.72
B	6.30	9.83	2.72
C	2.18	6.30	2.72
D	5.72	6.50	2.72
E	2.39	3.96	2.72
F	4.17	8.66	5.90
G	4.50	4.09	5.90
H	5.44	1.01	5.90
I	7.32	8.53	8.80
J	3.48	8.64	8.80
K	5.72	3.00	8.80
L	6.63	4.34	11.90
M	5.31	1.37	11.90

Table A.2. NIST oxygen cone-calorimeter tests fuel properties

Fuel package properties	Small triangle pallet	Large triangle pallet (1)	Large triangle pallet (2)
Mass of excelsior	7.4 kg (16.3 lb)	14.8 kg (32.6 lb)	14.8 kg (32.6 lb)
Mass of three pallets	27.6 kg (60.9 lb)	55.9 kg (123.2 lb)	53.9 kg (118.8 lb)
Total mass of fuel package	35 kg (77.2 lb)	70.7 kg (155.9 lb)	68.7 kg (151.4 lb)
Moisture content	5–24 %	6–11 %	7–12 %
Pallet dimensions	.94 m × .94 m × 0.089 m (3.0 ft × 3.0 ft × 0.3 ft)	1.22 m × 1.02 m × .13 m (4 ft × 3 ft 4 in × 5 in)	1.22 m × 1.02 m × .13 m (4 ft × 3 ft 4 in × 5 in)

Table A.3. Initial conditions for fire tests

| Test # | Fire | | Configuration | | | | Moisture content (%) | Amb temp. (°C) |
	Small (wood crib)	Large (pallet)	Door 2 Open	Door 2 Closed	Crib/pallet mass (g)	Excelsior mass (g)		
1	X		X		21,275	3977	5.4	−7
1 (repeat)	X		X		21,630	3885	5.4	−8
1 (repeat)	X		X		21,184	3769	5.2	−8
2	X			X	20,637	3905	5.3	−7
2 (repeat)	X			X	19,869	3723	5.6	−7
3 Fail		X	X		27,098	3700	13–18	−6
3		X	X		26,540	7703	13–18	−6
3 (repeat)		X	X		21,441	7546	12–19	−6
4		X		X	27,650	7463	13–17	−6
4 (repeat)		X		X	27,226	7556	10–18	−6

Table A.4. Event times for fire tests

Test #	Test start time	Free burn time	Occ sensor	Crib collapse	Ventilation
1	Wed, 1:20 pm	1:20	2:30	13:00	23
1 (repeat)	Thurs, 9:30 am	1:30	2:15, 8:30	14:30	20
1(repeat)	Thurs, 12:55 pm	1:20	3:40	13:10	20
2	Wed, 2:45 pm	1:25	2:30	13:00	23
2 (repeat)	Thurs, 10:45 am	1:15	2:30, 12:30	14:00	20
3 Fail	Thurs, 2:10 pm	NA	2:30	NA	11
3	Thurs, 2:35 pm	1:15	2:30	8:20	21
3 (repeat)	Friday, 8:00 am	1:20	3:40	10:15	20
4	Friday, 8:45 am	1:10	4:30, 6:30-2nd, 3rd	8:30	20
4 (repeat)	Friday 9:30 am	1:10	8:11, 1st, 2nd, 3rd	8:10	20

A.3   Multi-criteria Smoke Detector Description

The photoelectric sensor uses optical detection and can measure the specific optical density of smoke. This specific optical density is also called smoke obscuration. Obscuration is a unit of measurement that has become the standard definition of smoke detector sensitivity. This is a useful measurement that quantifies the concentration of smoke in the environment. The detector combines the smoke obscuration measurement with temperature and carbon monoxide measurements for enhanced fire detection (Figs. A.6, A.7, A.8, A.9, A.10, A.11, A.12, and A.13).

Model FDOOTC441 can measure smoke obscuration between 0 % and 3 %. This is a measurement of the smoke density present having the capacity of blocking a percent of a light source per linear foot. This is based on a system having a light source at one end and a light receiver at the other, typically 100 m apart, and

Fig. A.6. Oxygen cone-calorimeter test of the MFRI fuel package by NIST test

Fig. A.7. Various pallets
used for testing

the measurement is based on the percent of light lost with the smoke interrupts the
light. The detector can only measure up to 3 % smoke obscuration, after which it
goes into a smoke alarm state. This is a typical threshold for a smoke detector and
this limitation should be considered when trying use measured smoke density for
fire calculations. The reported smoke obscuration is the amount of smoke needed
to go into alarm, not the amount of smoke in the chamber. For example, a reported

Fig. A.8. Cut pallets used
for testing

Fig. A.9. Temperature and
smoke obscuration data
request

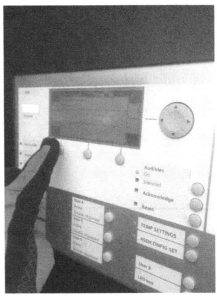

obscuration level of 3 % means that there is 0 % within the chamber and 3 % is
needed to go into alarm.

Model FDOOTC441 provides five field-selectable temperature thresholds,
ranging from 57 °C to 79 °C. As the conditions were expected to exceed all of the
thresholds, the temperature threshold chosen for the experiments was the highest
of 79 °C. A CO alarm response time is dependent on the CO concentration (parts

```
|..............................................................|
| Temperature report    Module  1/2                02/26/15 11:13:32 AM |
|                       FDnet card                              026 |
| Device  1/2/11/1      Not applicable      HTRI-D              |
| Device  1/2/11/3      Not applicable      HTRI-D              |
| Device  1/2/11/5      Not applicable      HTRI-D              |
| Device  1/2/11/7      Not applicable      HTRI-D              |
| Device  1/2/11/9      Not applicable      HTRI-D              |
| Device  1/2/11/11     Not applicable      HTRI-D              |
| Device  1/2/11/13     Not applicable      HTRI-D              |
| Device  1/2/11/15     Not applicable      HTRI-D              |
| Device  1/2/11/17     Not applicable      HTRI-D              |
| Device  1/2/11/19     Not applicable      HTRI-D              |
| Device  1/2/11/21     Not applicable      HTRI-D              |
| Device  1/2/11/23     Not applicable      HTRI-D              |
| Device  1/2/11/25     Not applicable      HTRI-D              |
| Device  1/2/11/27     Not applicable      HTRI-D              |
| Device  1/2/11/29     Not applicable      HTRI-D              |
| Device  1/2/11/30     42 ???F             SMOKE 30            |
| Device  1/2/11/31     63 ???F             SMOKE A ROOM 1      |
| Device  1/2/11/32     72 ???F             SMOKE B ROOM 1      |
| Device  1/2/21/36     36 ???F             SMOKE F ROOM 3      |
| Device  1/2/21/37     35 ???F             SMOKE G ROOM 3      |
| Device  1/2/21/38     34 ???F             SMOKE H ROOM 4      |
| Device  1/2/21/39     36 ???F             SMOKE I ROOM 6      |
| Device  1/2/21/40     37 ???F             SMOKE J ROOM 5      |
| Device  1/2/21/41     40 ???F             SMOKE K ROOM 7      |
| Device  1/2/21/42     37 ???F             SMOKE L ROOM 8      |
| Device  1/2/21/43     40 ???F             SMOKE P ROOM 9      |
```

Fig. A.10. Temperature data output from Siemens system

```
|..............................................................|
| Sensitivity report    Module  1/2                02/26/15 11:13:40 AM |
|                       FDnet card                              026 |
| Device  1/2/11/1      Not applicable      HTRI-D              |
| Device  1/2/11/3      Not applicable      HTRI-D              |
| Device  1/2/11/5      Not applicable      HTRI-D              |
| Device  1/2/11/7      Not applicable      HTRI-D              |
| Device  1/2/11/9      Not applicable      HTRI-D              |
| Device  1/2/11/11     Not applicable      HTRI-D              |
| Device  1/2/11/13     Not applicable      HTRI-D              |
| Device  1/2/11/15     Not applicable      HTRI-D              |
| Device  1/2/11/17     Not applicable      HTRI-D              |
| Device  1/2/11/19     Not applicable      HTRI-D              |
| Device  1/2/11/21     Not applicable      HTRI-D              |
| Device  1/2/11/23     Not applicable      HTRI-D              |
| Device  1/2/11/25     Not applicable      HTRI-D              |
| Device  1/2/11/27     Not applicable      HTRI-D              |
| Device  1/2/11/29     Not applicable      HTRI-D              |
| Device  1/2/11/30     2.44 %/ft           SMOKE 30            |
| Device  1/2/11/31     2.50 %/ft           SMOKE A ROOM 1      |
| Device  1/2/11/32     2.44 %/ft           SMOKE B ROOM 1      |
| Device  1/2/21/36     2.38 %/ft           SMOKE F ROOM 3      |
| Device  1/2/21/37     2.31 %/ft           SMOKE G ROOM 3      |
| Device  1/2/21/38     1.94 %/ft           SMOKE H ROOM 4      |
| Device  1/2/21/39     1.63 %/ft           SMOKE I ROOM 6      |
| Device  1/2/21/40     2.06 %/ft           SMOKE J ROOM 5      |
| Device  1/2/21/41     2.31 %/ft           SMOKE K ROOM 7      |
| Device  1/2/21/42     1.63 %/ft           SMOKE L ROOM 8      |
| Device  1/2/21/43     2.25 %/ft           SMOKE P ROOM 9      |
```

Fig. A.11. Smoke obscuration data output from Siemens system

```
|  ...... ....... ........ ........                02/26/15  11:04:49 AM |
| NEW Zone         3/1        Active     IN        02/26/15  11:04:49 AM |
|     01-02 HTRI-D SECURITY SWITCH                                       |
| NEW Zone         3/1        Supervisory IN       02/26/15  11:04:49 AM |
|     01-02 HTRI-D SECURITY SWITCH                                       |
| NEW Zone        16/1        Active     IN        02/26/15  11:04:51 AM |
|     15-01 HTRI-D SECURITY SWITCH                                       |
| NEW Zone        16/1        Supervisory IN       02/26/15  11:04:51 AM |
|     15-01 HTRI-D SECURITY SWITCH                                       |
| ACK Zone         3/1        Supervisory IN       02/26/15  11:04:53 AM |
|     01-02 HTRI-D SECURITY SWITCH                      from PMI    1    |
| ACK Zone        16/1        Supervisory IN       02/26/15  11:04:53 AM |
|     15-01 HTRI-D SECURITY SWITCH                      from PMI    1    |
|......................................................................|
```

Fig. A.12. Security switch data output from Siemens system

Fig. A.13. Siemens smoke detector file

per million, ppm). An advance warning can be issued when 30 ppm CO is exceeded, and a pre-alarm when 50 ppm CO is exceed. A full CO alarm will activate at 70 ppm CO within 60–240 min, 150 ppm CO within 10–50 min, and 400 ppm CO within 4–15 min. It is also noted in the product specifications that at temperatures greater than 79 °C the CO sensor may not function reliably. Since these conditions are expected, this limitation will be taken into account when analyzing the results.

A.4   Load Cell Limitation

The expected heat release rate of the pallet fire is too extreme and could possibly damage the load cell during the experiment. In order to gather as much information about the mass loss as possible, the mass of the fuel is weighed before and a photo of the leftover fuel is taken at the end of an experiment. This information is provided in the Appendix. It is recommended in future tests to measure the left over fuel to obtain the entire mass loss during the fire evolution.

A.5   Fuel Package Location

The larger fire compartments on the first floor (300 ft^2) as compared to the third and fourth floor smaller burn rooms about the size of 115 ft^2 are better for this project's goals. The larger fire room provides more ventilation for the fire and limits the possibility of extreme conditions within the fire room. In a larger burn compartment the temperatures will not reach as high as a smaller room with similar ventilation doors. In a larger fire compartment the fire is further away from the walls limiting the heat exposure from the flames as well. The newly replaced Padgenite fire resistant panels on the first floor will be able to better handle the heat than the older panels on the third and fourth floors. This project is unique in that the conditions throughout the entire building will be observed, rather than past experiments where one or two compartments are observed.

A.6   Wood Crib Fuel Description

The wood cribs are made of furring strip board. The length, size, and number of individual wood members and their arrangement in the crib are specified as follows for a classification and rating of 1-A; 72 wood members, each 38 mm by 38 mm by 500 mm in 12 layers of six members. The layers of the wood crib consist of specified sizes and lengths of furring strip board placed at right angles to one another. The UL Standard also prescribes the amount of and the heptane pan size. The amount of heptane required is 1.1 L within a 400 mm by 400 mm by 100 mm pan.

After determining the fuel set up, it was realized that tests within the MFRI structural fire fighting building are not allowed to use petroleum based products. Excelsior is a commonly used fuel source for training purposes in the structure. It is then considered to replace the heptane pool fire with excelsior to ignite the crib. Tests are conducted within the fire lab at the University of Maryland to determine if excelsior alone can ignite the wood crib and how much excelsior is needed. It is determined that about 3 kg of excelsior underneath the wood crib is enough to ignite the crib in about 1 min. This is a similar time as it took the heptane pool fire to ignite the crib.

A.7 Pallet Fuel Description

The typical larger triangle package of three full sized pallets and excelsior was also considered in this study. The heat release rate of the previous crib fuel package could be determined using the properties of wood, and information from past studies.

The mass of the pallets for the large triangle package was about 70 kg, and the smaller package was about 35 kg with the pallets weighing about 27.5 kg and the excelsior weighing about 7.5 kg. Flat and upright fuel arrangements were also tested by NIST but were not used for this study as they were not as representative of the typical fuel package used in normal training exercises. It was also considered to use only excelsior for the fuel, but it was concluded that the resulting fire environment would provide a fast growing fire that would result in high temperatures inside the burn room without enough time for the smoke to travel throughout the building.

The various fuel packages were tested by NIST and their heat release rates were determined. The fuel packages were arranged in a room corner composed of 2.44 m × 2.44 m (8' × 8') walls covered with two layers of 13 mm (0.5') gypsum board with a partial ceiling over the corner also made of gypsum board as seen in Fig. 2.14. The experiments were conducted under the oxygen consumption calorimeter at NIST to collect heat release data. One heat flux gauge at a height of 1 m (3'–3") was directed at the fuel package. The heat flux gauge was located 1 m (3'–3") from the right wall and 2.44 m (8') from the left wall.

The moisture content each fuel is also measured directly before each test to ensure that the exact moisture content is known for the experiment. The cribs are kept in a climate-controlled area from the time they are assembled to the time they are used for testing. The moisture content of the cribs is mostly uniform and ranged from 5.2 % to 5.6 %. The pallets are not kept in a controlled environment, and are just stored outside of the MFRI structural fire fighting building. During the experiments there was still snow on the ground and some of the pallets had snow residue. A photo of the pallets used is shown in the Appendix. The moisture content of the pallets varies and at least four measurements are taken for each pallet.

In order to have the same fuel load as the chosen NIST test, the pallets are cut in half, the cut pallets are shown in in the Appendix. For the NIST test, the combined weight of the pallets was 27 kg, and the excelsior was 7.5 kg. For each test, the three pallets are chosen that provide the closest combined weight of 27 kg. The 7.5 kg of excelsior is also measured for each test. The weights of the three pallet fuel load are noted in the Appendix.

A.8 Testing Environment Conditions

The tests were performed in February of 2015. It was one of the coldest weeks of the year, with the average temperature being about −6 °C (20 °F). There was snow on the ground every day which caused some delays. The area had to be prepared for the tests and the equipment through snow plowing and salting areas that would be used for walkways. As it was predicted to snow again on a few of the nights between each day's testing, a tent was provided to cover the control and monitoring system outside of the building. It was not feasible to take down the system every day, even though was a temporary system, because it took about a day to install. Inside the tent was the portable fire alarm control system, the thermocouple monitoring system, and a laptop that was gathering the information from the fire alarm control panel. The tent also provided storage for the other equipment used for installation and testing. The tent could house about six people which was useful as the test required someone to monitor each of the recording devices. Also during the test someone had to be in communication with the fire fighters inside the structure conducting the test. Harsh outside conditions affected the fire results, as described before but it also affected the installation and testing procedures. The extreme cold and snow resulted in many needed breaks inside the heated MFRI institute. More consideration had to be taken when locating things so that electrical equipment could be protected within the tent. The hard work of student volunteers, MFRI fire fighters, and Siemens employees all made this possible during the testing week.

A.8   Fire Scenario Descriptions

For each scenario, both detectors A and B go into smoke, temperature, and CO alarm. In each case, B goes into alarm just before A does. This makes sense as the door in front of detector A is closed and the door in front of detector B is open and is exposed to the elements of the fire first. The first time that the system is able to report temperature and smoke obscuration readings is displayed within the second blue square of Figs. 3.1 and 3.2. During the first few minutes of each test, the system is overloaded with sending alarm signals that the temperature and smoke obscuration measurements are not conveyed. This is important limitation of the system that was not expected.

The results of the crib fuel source were more in line with expectations than the results of the pallet fuel source. During the crib test, detectors A and B initiated alarms as well as some of the third and fourth floor smoke alarms. The events shown in purple for the crib test are the third and fourth floor detectors that go into a smoke alarm state. The locations of the K, M, I, and L detectors of the third and fourth floors are provided in the approach and their coordinates are in the Appendix. During the crib test, the contact sensors reported an open state after ventilation was initiated. The only sensor that didn't operate as planned is the occupancy sensor, which never signaled an activated state.

The pallet test shown in Figs. 3.2 and 3.4 produced different results than what was expected. The first alarm for the pallet test is temperature, and the second is the smoke. This could be due to the fast growth rate of the larger amount of excelsior that resulted in high temperatures within a short amount of time. Smoke alarms on the third and fourth floors are not activated during this test, which is surprising because the smoke obscuration measurements did reach the 3 % threshold. Also the contact sensors reported open at an early stage of the fire before ventilation was initiated.

For both the large and small experiment the first floor is the first to exceed the 3 % threshold, and the second floor is the last. This is because the stairway is an open grated stairwell that acts like a shaft and allows the smoke to travel upward from the first floor into the fourth and third floors first. Comparing these results with the timeline, it can be seen that more detectors reach the 3 % threshold than the number of detectors that reported a smoke alarm. For the large experiment most of the detectors exceed the 3 % threshold within the first 5 min of the experiment, and yet none of the detectors on the third or fourth floors reported a smoke alarm.

Temperatures recorded by detector B are higher than the temperature recorded by detector A. This makes sense as the door in front of the A detector is closed and the door in front of the B detector is open and is closer to the fire. The temperatures for detectors A and B for the crib fire stay elevated for a much longer time than for the pallet fire. The crib fire has a longer steady heat release rate that cause the temperatures to stay elevated, whereas the pallet fire grew very large quickly but also decayed quickly. The temperatures on the upper floors were slightly elevated during the tests. The third and fourth floors exhibit the most temperature rise of the upper floors. The open stair behaving like a shaft explains this trend.

Bibliography

1. "Trends and Patterns of U.S. Fire Losses in 2012," National Fire Protection Association Fire Analysis and Research Division, 2012, http://www.nfpa.org/~/media/Files/Research/NFPA%20reports/Overall%20Fire %20Statistics/ostrends.pdf.
2. Aram Kalousdian, "Smart Buildings and Smart Communities Bring Measurable Results," CISCO, 2014, http://www.sustainablecitynetwork.com/topic_channels/building_housing/article _151bba4c-c913-11e0-a9f7-0019bb30f31a.html.
3. J. Lataille, "Factors in Performance-Based Design of Facility Fire Protection," Society of Fire Protection Engineering, 2008, http://magazine.sfpe.org/fireprotection-design/factors-performance-based-design-facility-fire-protection.
4. "National BIM Standard. United States Version," National Institute of Building Sciences buildingSMART alliance, http://www.nationalbimstandard.org/.
5. M. Gallaher, "Cost Analysis of Inadequate Interoperability in the U.S. Capital Facilities Industry," 2004, http://fire.nist.gov/bfrlpubs/build04/art022.html.
6. Bentley, AECOsim Building Designer V8i. Multi-discipline Building Information Modeling (BIM) Software, 2014, http://www.bentley.com/enUS/Products/AECOsim+Building+Designer/AECOsim+Building+Designer+Structural+Engineers+Software+Features.htm.
7. National Institute of Standards and Technology, Fire Dynamics Simulator, Version 6.0, http://www.fire.nist.gov/fds/.
8. W.S. Lee, and S.K. Lee, "The estimation of fire location and heat release rate by using sequential inverse method," Journal of the Chinese Society of Mechanical Engineers. (2005) Vol 26, No. 2. pp. 201–207.
9. W.D. Davis, and G.P. Forney, "A sensor-driven fire model version 1.1," National Institute of Standards and Technology, NISTIR 6705, 2001.
10. W.D. Davis. "A sensor-driven fire model version 1.2," National Institute of Standards and Technology, NIST SP 1110, Jul 2010.
11. W. Jahn, G. Rein, J. Torero, "Data assimilation in enclosure fire dynamics – towards adjoin modeling." EUROGEN, Cracow, Poland, 2009, http://www.see.ed.ac.uk/~grein/rein_papers/Jahn_Eurogen2009.pdf.
12. L. Han, et al, "FireGrid: An e-infrastructure for next-generation emergency response support," Volume 70, Issue 11, November 2010, Pages 1128–1141. doi:10.1016/j.jpdc.2010.06.005.
13. A. Neviackas, A. Trouve, "Inverse Fire Modeling To Estimate, The Heat Release Rate Of Compartment Fires," University of Maryland, Department of Fire Protection Engineering, 2007.
14. M. Price, "Using Inverse Fire Modeling With Multiple Input Signals to Obtain Heat Release Rates in Compartment Fire Scenarios," University of Maryland, Department of Fire Protection Engineering, 2014.

© The Author(s) 2016
R.F. Wills, A. Marshall, *Development of a Cyber Physical System for Fire Safety*, SpringerBriefs in Fire, DOI 10.1007/978-3-319-47124-2

15. A. Anderson, O. Ezekoye, "Property Risk Optimization by Predictive Hazard Evaluation Tool (PROPHET)," The University of Texas at Austin, 2014.
16. "SOCIETY OF FIRE PROTECTION ENGINEERS POSITION STATEMENT P-05-11," Building Information Modeling and Fire Protection Engineering. October 23, 2011. http://www.sfpe.org/Portals/sfpepub/Prof%20Prac/111023%20SFPE%20BIM0POSITION%20STATEMENT%20Final.pdf.
17. W. Jones et al., "Workshop to Define Information Needed by Emergency Responders During Building Emergencies," National Institute of Standards and Technologies Building and Fire Publications, NISTIR 7193, 2005.
18. D. Holmberg, W. Davis, S.Treado, K. Reed, "Building Tactical Information System for Public Safety Officials, Intelligent Building Response (iBR)," National Institute of Standards and Technologies Building and Fire Publications, NISTIR 7314, 2006.
19. A. Tewarson, "Generation of heat and gaseous, liquid, and solid products in fires," The SFPE Handbook of Fire Protection Engineering, 4th Edn. SFPE and NFPA. Quincy, MA, 2008.
20. M, Stolp, "The extinction of small wood crib fires by water," 5th Int. Fire Protection Senminar, Karlsruhe. Verinigung zur Forderung des Deutchen Brandshutzes – VFDB – 1976: 127–41.
21. Underwriters Laboratories Inc. "UL 711 Standard for Safety, Rating of Fire Testing of Fire Extinguishers," 2009. file:///Users/rosaliewills/Downloads/0711_7.pdf.
22. B. Jacobs, "Suppression Effectiveness of Water Sprays on Accelerated Wood Crib Fires," University of Maryland, Department of Fire Protection Engineering, 2011.
23. D. Madrzykowski, "Fatal Training Fires: Fire Analysis for the Fire Service," Building and Fire Research Laboratory, National Institute of Standards and Technology, 2008.
24. ASTM E1354, "Standard Test Method for Heat and Visible Smoke Release Rates for Materials and Producers Using an Oxygen Consumption Calorimeter," ASTM International, 2014.
25. C. Lannon, J. Milke, "Evaluation of Fire Service Training Fires," Department of Fire Protection Engineering, University of Maryland, 2013.

Printed in the United States
By Bookmasters